D1254168

# BUILDING COMPETENCES IN THE FIRM

# Building Competences in the Firm

## Lessons from Japanese and European Optoelectronics

Kumiko Miyazaki

Foreword by Christopher Freeman

**St. Martin's Press**

First published in Great Britain 1995 by
MACMILLAN PRESS LTD
Houndmills, Basingstoke, Hampshire RG21 2XS
and London
Companies and representatives
throughout the world

A catalogue record for this book is available
from the British Library.

ISBN 0–333–61673–1

| 10 | 9  | 8  | 7  | 6  | 5  | 4  | 3  | 2  | 1  |
|----|----|----|----|----|----|----|----|----|----|
| 04 | 03 | 02 | 01 | 00 | 99 | 98 | 97 | 96 | 95 |

Printed and bound in Great Britain by
Antony Rowe Ltd
Chippenham, Wiltshire

First published in the United States of America 1995 by
Scholarly and Reference Division,
ST. MARTIN'S PRESS, INC.,
175 Fifth Avenue,
New York, N.Y. 10010

ISBN 0–312–12314–0

Library of Congress Cataloging-in-Publication Data
Miyazaki, Kumiko.
Building competences in the firm : lessons from Japanese and
European optoelectronics / Kumiko Miyazaki.
p. cm.
Includes bibliographical references and index.
ISBN 0–312–12314–0
1. Technological innovations—Management.   2. Optoelectronics
industry—Japan.   3. Optoelectronics industry—Europe.
4. Employees—Effect of technological innovations on—Japan.
5. Employees—Effect of technological innovations on—Europe.
I. Title.
HD45.M55  1995
338.4'7621381045'094—dc20                        94–20757
                                                      CIP

# Contents

# List of Figures

# List of Tables

# Foreword

This is a book of outstanding interest for several reasons. In the first place, it is universally agreed today that information and communication technology (ICT) has become the dominant technology in the world economy. Optoelectronics has now become a leading core technology within ICT, comparable to microelectronics in its significance.

Secondly, as Kumiko Miyazaki shows, optoelectronics differs from mechatronics in one fundamental respect – whereas mechatronics is an example of 'production fusion' across several industrial sectors, optoelectronics is a case of 'scientific fusion'. Consequently the problems of diversification for firms in the ICT industry, who wished to enter the optoelectronics field, differed fundamentally from those who diversified across industrial sectors.

Thirdly, this type of technological diversification confronts firms with very different problems of management strategy, involving the development of competence in science as well as technology. Kumiko Miyazaki's account of this process of 'competence building' is of absorbing interest for students of business strategy, as well as for scientists and technologists.

Fourthly, the book demonstrates that the process of building competence and sustaining it permits a change of technological trajectory so that firms can achieve much greater flexibility. This flexibility has become more and more important in sustaining international competitiveness.

Fifthly, the author uses novel techniques in her analysis of bibliometrics and patents statistics to explore the relative strengths and weaknesses of the various firms involved. This too is of the greatest interest to those working in the field of science and technology policy, as well as to those in business studies and strategic management.

Finally, since many of the leading firms in this technology are Japanese (for example, NEC, Fujitsu, Sumitomo, Sony, Sharp, Hitachi and Toshiba) the comparison of their performance with European firms such as Philips, GEC, Siemens and STC, provides an excellent opportunity to assess some of the characteristics of

Japanese management style. These characteristics continue to preoccupy economists and management students world-wide in their atempts to develop the theory of the firm.

For all these reasons I commend this book most strongly.

CHRISTOPHER FREEMAN
*Emeritus Professor of Science and Technology Policy*
*University of Sussex*

# Acknowledgements

This book attempts to explore the roots of competitiveness of firms. I became interested in the notion of core competence just after I obtained my MBA at INSEAD, where I stayed on to work as a researcher. I wanted to explore how firms assimilate and exploit radically new technologies to their advantage. I came to University of Sussex (Science Policy Research Unit) to start my studies for a doctorate in 1989. During the next few years, I applied my background in several areas, including my physics degree, industrial experience, business administration, and being bi-lingual, to embark on this project. SPRU offered an intellectually stimulating, friendly environment.

My supervisor at Sussex University, Keith Pavitt, always challenged me with new ideas and gave constructive criticisms and supported me. I am equally grateful to my co-supervisor Roy Rothwell, for his essential guidance. Under these supervisors, it was possible to make progress along paths which evolved through the merger of two complementary streams of thoughts and techniques.

I would like to acknowledge the contribution of many people at SPRU (too many to mention) through numerous discussions. Nick von Tunzelmann has been a valuable source of comments, and gave me advice on statistical analysis and on my drafts. Chris Freeman has been a source of inspiration and his encouragement and kindness were much appreciated. Discussions with Mike Hobday were stimulating and helpful. In the economics department, thanks must go to Peter Holmes for his helpful comments on the theoretical chapter. I also acknowledge the constructive suggestions from former colleagues at INSEAD during the initial stage of my research.

I am grateful to Professor Kodama for welcoming me to join his group at NISTEP (National Institute of Science and Technology Policy) when I did my fieldwork in Japan. It was both a very productive and very pleasant experience. Discussions with Dr Tsukahara were also helpful. I am very grateful to many people in the companies and institutions who kindly cooperated with my research.

Thanks must go to SPRU and the Designated Research Centre at SPRU for funding my trips to attend some international conferences and for providing some funds to carry out the INSPEC database search. I am also grateful to SPRU for providing me with some funding for converting my D.Phil thesis into this book. Mike Gibbons,

xii

Chris Freeman, Nick von Tunzelmann, Ben Martin, Peter Holmes, John Surrey and Rod Coombs read drafts or parts of this book. The final version takes into account their valuable comments. I would also like to thank Chris Freeman for kindly writing the Foreword.

The author and publishers wish to thank the following for permission to reproduce copyright material:

McGraw-Hill, for Figure 2.2, from M. Giger, *Les Bonzais de l'Industrie Japonais* (1984).

OECD, for Table 3.3, from *Technology Fusion: a Path to Innovation: the Case of Optoelectronics* (1993).

National Academy Press, for Figure 2.4, from S. J. Kline and N. Rosenberg. 'An Overview of Innovation', in R. Landau and N. Rosenberg (eds). *The Positive Sumstrategy: Harnessing Technology for Economic Growth* (1986).

Yasinori Baba, for Figure 2.5, from 'Systemic Innovation: Its Nature and How to Benefit from it', University of Sussex, SPRU, Mimeo (1988).

OITDA, for Figure 3.1 and Table 3.1 and Appendix B.1 and B.2.

Fraunhofer Institute for Systems and Innovation Research (FhG-ISI) for Figure 5.1, from H. Grupp, *Technology Indicators in Corporate Forecasting* (1989).

Hitachi Research Institute, for the diagram of a systemic concept in Appendix A.

Hitachi Research Laboratory, for the diagram of the trend of liquid crystal displays in Appendix C.

HMSO, for Table 3.2, from *Optoelectronics: Building on Our Investment* (1988).

STN International, for the INSPEC record data in Appendix F.

The collaboration of OTAF, US Department of Commerce, in compiling the US patent data is gratefully acknowledged. I would like to express my gratitude to SPRU secretaries, librarians and other administrative staff for their help. Finally, special thanks go to my parents for their encouragement and support.

KUMIKO MIYAZAKI

# List of Abbreviations

| | |
|---|---|
| ACOST | Advisory Council on Science and Technology |
| ALE | Atomic Layer Epitaxy |
| AT&T | American Telephone and Telegraph Company |
| AV | Audio Visual |
| BT | British Telecommunications |
| C&C | Computers and Communications |
| CCD | Charge Coupled Device |
| CD | Compact Disk |
| CD-I | Compact Disk Interactive |
| CD-ROM | Compact Disk Read only memory |
| CMOS | Complementary Metal Oxide Semiconductor |
| COG | Chip on Glass |
| CRT | Cathode Ray Tube |
| CVD | Chemical Vapour Deposition |
| CT | Controlled Terms (INSPEC) |
| DFB | Distributed Feedback (laser) |
| DINA | Distributed Information Processing Network Architecture |
| DRAM | Dynamic Random Access Memory |
| DSM | Direct Scattering Mode |
| EC | European Community |
| E&E | Energy and Electronics |
| ESPRIT | European Strategic Programme of Research in Information Technology |
| ETL | Electro-Technical Laboratory (of Japan) |
| FDM | Frequency Division Multiplexing |
| FOS | The Fibreoptic and Optoelectronic Scheme |
| GaAs | Gallium Arsenide |
| GaAlAs | Gallium Aluminium Arsenide |
| Ge | Germanium |
| GEC | General Electric Company |
| HEMT | High Electron Mobility Transfer |
| HIOVIS | Highly Interactive Optical Visual Information System |
| IBM | International Business Machines |
| IC | Integrated Circuit |
| ICOT | Institute for the New Generation Computer Technology |
| IEE | Institution of Electrical Engineers |
| IFTI | Intra-firm Technology Indices |

| | |
|---|---|
| IFTI$_{INS}$ | IFTI based on INSPEC |
| IFTI$_{Pat}$ | IFTI based on patent data |
| InGaAsP | Indium Gallium Arsenide Phosphide |
| InP | Indium Phosphide |
| ISDN | Integrated Services Digital Network |
| IT | Information Technology |
| ITT | International Telephone and Telegraph |
| JESSI | Joint European Submicron Silicon Initiative |
| JOERS | Joint Optoelectronics Research Scheme |
| JVC | Victor Company of Japan |
| kbit/s | Kilo bits per second |
| LAN | Local Area Network |
| LCD | Liquid Crystal Display |
| LED | Light Emitting Diode |
| LPE | Liquid Phase Epitaxy |
| MBE | Molecular Beam Epitaxy |
| Mbit/s | Mega bits per second |
| MCVD | Modified Chemical Vapour Deposition |
| MOCVD | Metal Organic Chemical Vapour Deposition |
| MoD | Ministry of Defence |
| MITI | Ministry of International Trade and Industry (of Japan) |
| MQW | Multi-quantum Well (laser) |
| NEC | Nippon Electric Company |
| NTT | Nippon Telephone and Telegraph Company (of Japan) |
| OA | Office Automation |
| OECD | Organization for Economic Cooperation and Development |
| OEIC | Optoelectronic Integrated Circuit |
| OEM | Original Equipment Manufacturer |
| OITDA | Optoelectronic Industry and Technology Development Association (of Japan) |
| OPGW | Optical Fibre Composite Aerial Ground Wire |
| OTAF | Office of Technology Assessment and Forecast |
| OVPO | Outside Vapour Phase Oxidization Process |
| PBX | Private Branch Exchange |
| PC | Personal Computer |
| PCM | Pulse Code Modulation |
| PIN-FET | P-type Intrinsic N-type Field Effect Transistor |
| PIN-PD | P-type Intrinsic N-type Photodiode |
| PTT | Post, Telephone and Telegraph Administration |
| RACE | Research for Advanced Communications in Europe |

| | |
|---|---|
| RSRE | Royal Signals and Radar Establishment |
| RTA | Revealed Technology Advantage |
| SBU | Strategic Business Unit |
| SEM | Scanning Electron Microscopy |
| SPSS | Statistical Package for the Social Science |
| ST | Supplementary Terms (INSPEC) |
| STC | Standard Telephones and Cables Company Ltd |
| STL | Standard Telephones and Cables Laboratory |
| STN | Super-twisted Nematic (LCD) |
| TAB | Tape Automated Bonding |
| TEM | Transmission Electron Microscopy |
| TFT | Thin Film Transistor |
| III-V | semiconductor – A compound composed of elements from groups III and V of the periodic table |
| TN | Twisted Nematic (LCD) |
| Tokken | Abbreviation of Tokubetsu Kenkyu (Top priority project, Japanese) |
| VAD | Vapour-phase Axial Deposition |
| VCR | Video Casette Recorder |
| VLSI | Very Large-scale Integrated Circuit |
| VSIS | V-Channelled Substrate Inner Stripe (laser) |
| VTR | Video Tape Recorder |
| WDM | Wavelength Division Multiplexing |
| WORM | Write Once Read Many Times |

# 1 Introduction

In recent years, the notion of competence, which embodies intangible assets, skills and creative resources accumulated over time in firms, has been widely recognized and discussed.[1] There has been a surge of interest in the idea of core competence. It has become the focus of attention not only among academics, but also among business consultants, journalists, government officials and indeed business leaders.[2] Eastman Kodak, for example, has set themselves the goal of becoming 'the world leader in imaging'.[3] In order to achieve this goal, they have to be competitive in several functional areas, such as (1) customer focus, (2) manufacturing, (3) technology transfer, (4) reduction in product-cycle time. The separate business units which are in charge of a portfolio of apparently quite distinct imaging products are based on a cluster of competences. At Eastman Kodak, a number of core technologies have been identified, including optomechatronics, imaging science, imaging electronics, Xray-imaging material, silver halide and precision thin-film coating. The company recently launched a new product, a CD photography system, merging their traditional capabilities in silver halide with electronics.

While business leaders teamed up with business consultants to investigate their core competences at the firm level, the work of several academics[4] and other analysts has greatly enriched our understanding of firm-specific competences. These authors have argued that what firms have been able to do in the past strongly affects what they can hope to do in the future. Pavitt points out that given the cumulative nature of development and its uncertainty, the improvement of these competences requires continuous and collective learning. The notion of core competence suggests that there are firm-specific competences which are closely linked to a firm's intangible assets and its accumulated technological resources. In the resource-based perspective put forward by Teece, Pisano and Schuen (1990), firms are seen as heterogeneous, being different from one another with respect to their capabilities and endowments. This inter-firm difference forms the basis of competition. However, these analytical concepts have rarely been employed to provide empirical data. There is little in the way of formal methodology to approach the subject. The main purpose of this book is to provide an empirical analysis of the dynamics of competence building at the level of the firm. The development of empirical measures to assess the dynamics of competence building is crucially important for both management practice and the modern theory of the firm. This book casts original empirical light on the dynamics of competence-building processes in an international sample of companies engaged in optoelectronics,[5] which is a technological domain of growing importance, yielding a broad range of applications in several key sectors of the economy.

Much of the conventional literature on competence[6] has centred on first, picking a company 'A' which has been performing exceptionally well in recent years, and picking another company 'B' in the same industry which used to be well positioned but has gradually lost leadership to the former. The main part of the analysis centres on finding out the reasons behind the remarkable achievement of the winner, company 'A', compared to the flagging performance of the loser, company 'B'. The argument usually runs as follows: company 'A' was able to become a world leader because the top management was able to think in terms of building core competences, while company 'B' could not. The analysis then proceeds to examine what the company's core competences must have been to enable their transformation. In other words, the prevailing literature on competence analyzes firms on an *ex post* basis. In this book, I argue that it is possible to measure competence independent of company performance; in other words, *ex ante*.[7] Indeed, a novel technique to measure competence is introduced. Competence building can be traced dynamically so that it may be possible to predict a company's performance, to some extent.[8] Such a technique should be of interest not only to academics, but a wide range of people including business executives, corporate planners, R&D managers, investment analysts, consultants and policy makers.

I consider that there are several types of competence, such as financial, marketing, manufacturing, and technological. Core competence then, involves the integration of these sub-areas of competences. It is our understanding that these different areas of competence may initially be treated independently, in conducting the analysis. The technology component is enormously important, playing a strategic role in affecting the competitiveness of a firm. Thus, while we are fully aware that other forms of competence may be relevant in determining the core competence, we will be focusing on technological (R&D) competence, since it may be a necessary though unlikely to be a sufficient condition for a firm to be competitive. A company that has a strong technological position should be expected to be at least one that will have a more than average chance of being profitable.[9] Technological competence is a firm's change-generating capacity in regard to technologies. It encompasses a package of skills enabling the firm to respond to technological opportunities and to assimilate them into their existing core capabilities. Core competence is the ability to deploy and especially expand the range and implications of core capabilities. It is also likely to involve the integration of various streams of technologies and effective mobilization of resources across firms. We have focused on this area not only because of its importance, but because it has been a neglected area of research. Much of the work on the competitiveness of Japanese firms has focused on their manufacturing competence. Little is known of inter-firm differences with regard to the dynamics of technological competence building.

There is a tendency in the West to consider Japanese firms as being similar to each other but distinct from Western companies.[10] We tend to think that Japanese companies are managed by a special style of management. Less attention is given to how Toshiba differs from Hitachi, or NEC differs from Sony or Fujitsu.

This book traces the competence-building processes of firms by focusing on optoelectronics. Analysis of the evolution of firms by applying the notion of competence building in the same domain highlights inter-firm differences. How did firms' entry strategies differ from one another? What were the factors which affected competence building? How did they exploit the technology to their advantage? For example, it is found that Sony's competence centred on optical disk technologies including the key component, semiconductor lasers. Sony has also been strong in CCDs (Charge Coupled Devices) which are used in video cameras. Since the mid-1980s, Sony's optoelectronics competence building reflected the company's strategy of AV&CCC, focusing on Audio Visual & Computer, Communications and Components. On the other hand their competences in the areas of optical communications and liquid crystal displays have been weak. Siemens' analysis conversely showed its strength in optical communications-related areas. It was however weak in optical disks and liquid crystal displays (LCDS).

There were differences in the way optoelectronics competences accumulated even between Toshiba and Hitachi which are generally considered to be rival firms, similar to one another. Toshiba's technological paths show that their competence building was closely related to their corporate strategy of triple 'I', 'Information,

Intelligence and Integration'. Their strengths focused in areas related to IT (Information Technology), and not communications. In the 1960s, they entered optoelectronics hoping to develop a large memory device for information processing. Subsequently, they developed an optical disk filing system, and have maintained a leading position. Toshiba's entry into optoelectronics shows that they wanted to use this revolutionary technology in the field of IT which was perceived as an area of future growth. They were able to grasp this technological opportunity to strengthen what has now become their core business. At NEC, competence building has been closely related to optical communications. In recent years, they have been able to exploit their competences built through communications to strengthen other areas such as office products and consumer electronics applications. The analysis below shows NEC's relative weakness in optical disks. In liquid crystal displays, the company was a late starter, but has been able to catch up. Hitachi, on the other hand, was the only company whose optoelectronics competence was evenly balanced between optical communications and consumer industrial applications. In related fashion, the trajectories[11] of GEC demonstrated that their competence building was strongly related to their core capabilities in defence electronics and communications.

The analysis of the areas of strength and weakness in optoelectronics conducted below led to arranging the 11 firms into three groups; (1) firms whose optoelectronics-related competence building was driven by optical communications; (2) firms whose optoelectronics-related competence building was driven by industrial and consumer electronics; and (3) evenly balanced firms.

We identify the reasons behind the emerging strengths and weaknesses within optoelectronics for each firm. The observable difference in the competence-building process discussed above points to the following: Technological competence building is a cumulative process, where path dependence matters. In other words, what firms can do depends on a number of factors which are related to their past history and previous accomplishments.

We therefore argue that competence building is not a random process, but is constrained by two forces: historical, path dependent elements, such as top management strategy, primary markets and accumulated technological bases; and elements related to current organizational routines.[12] The emerging technological trajectories would be strongly dependent on the firm's core businesses and technological interests. Understanding and support by top management is essential for building competences. Competence building needs to be seen in the light of the overall strategy formulated by top management in the firm. Evidence suggests that competence building can be planned strategically, one step at a time, over a long period. The organizations can be changed gradually to ones more conducive to building competences. We shall show how some Japanese firms in particular, were able to take a long-term view towards accomplishing such organizational changes.

By contrast, neoclassical economics considers technology to have many of the properties of a public good, in a perfect competitive market.[13] Firms can easily acquire it from a pool of technologies available. Within such a framework, the strategy of firms would be to find an optimum solution for assimilating a radically new technology and we would not expect to see such diverging patterns. Our results are very different from this perspective. We have found neo-Schumpeterian theory a satisfactory framework for understanding competence building, since the model is dynamic, inter-firm difference exists and the model assumes the uncertain, groping nature of technological development.

The approach adopted here also examines how competences were altered by the long-term actions of the firms. Attention is drawn to the neglected area of the interlinkages between systems, key components and component generic technologies (the latter are used to develop the key components but are spread among several components so that they are generic in nature). In this study, we apply a three-level model, where systems and products at the highest level are composed of key components and other technologies. A key component gives high value added to the final products and is distinguished from 'commodity' components. Is there a common area in which all firms would be actively building capabilities? This book argues that firms make long-term investment to build technological capabilities in component generic technologies,[14] which in turn provide the foundation to develop key components. These are used in conjunction with other generic technologies to develop systems and products which depend on the firm's technological and existing business interests. This approach may offer companies strategic advantages over firms that opt for a policy of buying in key components. One of the main arguments of this book is that an effective vertical linkage between systems, key components and component generic technologies allows firms to capitalize on economies of scope, enabling them to diversify horizontally into new, and often unexpected areas. We therefore emphasize path dependent factors on the one hand. On the other hand we show that competence building can be steered in directions allowing firms to shift their trajectories depending on their proximity to the core capabilities. We also examine how search trajectories evolve during competence building. As we intend to show, competence building is an extremely long, arduous process, involving trial and error and continuous learning. It is very far from the kind of picking technologies 'off the shelf' portrayed by neoclassical economics.

The dynamic nature of competence building needs to be emphasized. Some firms may have been early starters, but may then fail to sustain competences. Others may have been late comers, but may then catch up, through various organizational routines. In this respect, our evidence suggests that some Japanese firms may have developed better procedures for building and sustaining competences.

From the point of view of an investigation into competence building, the choice of optoelectronics as a technological domain to study was based on several factors.

1. Optoelectronics consists of a new set of techniques, able to perform new functions beyond the scope of conventional electronic technologies. For example, fibre-optic cables have led to intercontinental optical communications, greatly expanding capacity and reducing costs. Bar code readers not only alleviate the work of cashiers, but improve stock control. Optical storage has led to audio compact disks, and CD-ROM, 'compact disk–read only memory' systems finding applications such as electronic publishing and online databases. Optoelectronics can be classified as a technological discontinuity. It is also a case of 'scientific fusion' and should be considered differently from the case of 'production fusion' as in mechatronics.[15] The choice of this field enables us to study how firms assimilate a radical emerging technology. It is a strategically important core technology, which must be mastered by firms in a variety of sectors in order to remain competitive. Firms have to choose which sub-areas of optoelectronics to enter in order to strengthen their competitive position.

2. It offers the possibility of understanding the interlinkages between systems, key components, and component generic technologies. Optoelectronics is an area where the three-level model proposed fits well. For example, an optical communication

system requires a light emitting component (e.g. a semiconductor laser), a light transmission component (e.g. optical fibre) and a light receiving component (e.g. a photodiode). These components have to work together. Innovation in one component affects others, just as new development in optical transmission affects the light emitting and receiving elements. These elements in turn are developed by a range of component generic technologies which are common to a number of components. Multilayer formations found in key components such as semiconductor lasers, photodiodes and light emitting diodes can be fabricated by common epitaxial growth techniques. Other generic component technologies include etching, life testing and transmission electron microscopy. A key component such as a semiconductor laser can be used in other applications including compact disks, by changing the performance parameters. The level of sophistication[16] of semiconductor lasers varies enormously depending on the application, as reflected in the price.[17] Firms which have developed competence in (say) semiconductor lasers for one application have the choice of entering new product markets via exploiting economies of scope.

3. It is a field where it is possible to see a clear pattern of innovation, from laboratory discovery in the 1960s, to a decade of uncertainty in the 1970s leading to successful commercialization and rapid diffusion in the 1980s.

4. It is a field characterized by fierce international competition, in which Japanese firms have caught up with or in some respects overtaken their US counterparts. The analysis of US patents showed that US and Japan accounted for approximately 70–80% of the total patenting in 1963–90 in optoelectronics. A comparison of patents in recent years to the whole period suggests that Japan has gained in relation to the US.[18] Thus we shall be analyzing the competence-building strategies of Japanese and European firms in an area where Japan is at the world frontier.

## 1.1  LIMITATIONS OF THE STUDY

A major point concerns the difficulty in gathering data. Since the area chosen touches on some delicate issues concerning the competitiveness of firms, some firms have been more cooperative than others. Most

Japanese firms and the two UK firms have been very cooperative. The continental firms, unfortunately, were less helpful.[19] The companies which cooperated were eager to give qualitative data on competence-building processes. However, data on optoelectronics-related R&D expenditures, product sales, market shares, number of scientists and so on were almost impossible to obtain. This was mainly because of the pervasive nature of optoelectronics, which finds applications in many product markets. This makes it difficult for the companies themselves to compile such data. In addition, even when such data existed, their confidential nature made the firms reluctant to disclose the information. Therefore, we have used publicly available data. In this study, an original technique has been developed to assess technological competence, in which US patenting and scientific publications are combined with interview data with the firm's scientists.

Following the point above, we would like to emphasize that this is not primarily a study about the optoelectronics industry. The sample of 11 firms chosen does not form a representative cross section of the industry. For example, it might be asked why we have not included any medium or small specialized firms which are actively involved in optoelectronics. If we were conducting a study on the industry, obviously it would have been necessary to include such firms. However, we are examining the competence-building strategies of large, multi-product, multi-technology firms, by taking optoelectronics as an example. The firms chosen had to be comparable in scale and scope to make comparisons. Also, because we were going to use publicly available data, the firms had to have adequate amounts of publications and US patents. Medium or small specialized firms would not have met the above criteria. Furthermore, of the many large firms that could have been chosen, the selection criteria depended on access and location.[20] The choice of 11 firms allowed me to conduct intensive interviews and at the same time apply statistical techniques. If it had been fewer than 11, it would have made statistical analysis difficult. If it had been more than 11, we would not have been able to conduct intensive interviews with each firm.

We would also like to underline that in this study, we are assessing the dynamics of competence building which are internal to the firm. The technique developed allows us to examine how competences are accumulated in the sub-fields of each firm. We are not comparing say, the competence in field 'i' of firms A, B and C. Rather, we are trying to understand why firm A has become more competent in field 'i' than fields 'j' or 'k'. Nor are we able to relate internal competences to the actual performance of firms. As we discussed earlier, a firm's competitiveness depends on an integration of competences such as technological, marketing and financing. In this study, we are focusing on one element, the technological, which we nevertheless believe to form the foundation of a firm's competitiveness.

Another point to note is that this study is not about national systems of innovation. Although we shall be examining 11 companies from three countries,[21] it is not our goal to make international comparisons. For this reason, we do not treat Japanese companies as a distinct group. For example, no reference is made to those aspects of Japanese management, such as life-time employment, or the company-based trade union system, which might affect the eagerness of Japanese firms to assimilate new technologies. Also, Japanese industries are believed to have benefited from the special ties they have with the financial institutions, which allowed them to pursue long-term strategies. It is widely held that the growth of Japanese industries in the 1950s and 1960s was fuelled by funds available from the Japan Development Bank.[22] For example, readily available low-interest funds from the Japan Development Bank and from commercial banks have facilitated the diffusion of robots in the 1980s.[23] Throughout the 1980s, cheap finance was readily available from banks and the stock market. We are fully aware that Japanese industries have benefited from these types of funding and that the cross-shareholding system has led to a stable industrial structure compared to the situation in Europe where mergers and acquisitions are common phenomena. A limitation of this study is that it does not examine the affect of financial institutions in the competence-building process. In summarizing this section, the approach taken in this study is the following: to begin with, we shall assume that Japanese firms are no different from European firms. Rather, in examining the competence-building processes, when we note specific national differences, we will draw attention to them.

## 1.2 STRUCTURE OF THE BOOK

The first aim of this study is to discuss the notion of competence and to develop a methodology for operationalizing it. We shall be showing how the concept of competence can be applied in real life to trace the evolution of firms. Such a scheme would enable us to understand the technology strategies of different firms. We will explore the potential of a technique combining three kinds of data (US patents, scientific publications and qualitative interview data) to assess the dynamics of competence building in optoelectronics within firms. At the same time we will examine the factors affecting the rate and direction of the development of competences in companies.

In the following chapters, we shall draw upon various elements of neo-Schumpeterian theory and the firm-specific competence approach, to examine the competence-building process of 11 firms actively engaged in the field of optoelectronics.

Chapter 2 develops an analytical framework. The main concepts involved in the notion of competence are discussed. We shall give definitions of the terms used in this study. A model of competence building is put forward, identifying factors affecting the rate and direction of competence building.

Chapter 3 presents an overview of the evolution of optoelectronics and its technical and economic features. The non-scientist unconcerned with technical details can skim through this chapter to proceed to Chapter 4, which examines the factors affecting the rate and direction of the competence-building process. They relate to both path-dependent (historical) and current factors, such as organizational routines. This chapter also attempts to summarize the model of the competence-building process. In Chapter 5 we outline the methodology to assess competences using the three types of data. Strengths and weaknesses of the method are discussed. Various statistical analyses are presented to examine the validity of the three measures. We shall extend the analysis to measure competences of the firms in the various sub-areas of optoelectronics. The areas of strength and weakness lead to a grouping of the firms into three categories. Those who are less keen on quantification can read sections 5.1, 5.4, 5.5 and 5.7.

Chapter 6 examines the linkages between systems, key components and component generic technologies through two methods: statistical techniques and technological linkage maps. The concepts of 'search zones' and incremental learning are also examined. While competence building is constrained by path dependence, once competence building reaches a certain point, it provides the opportunity for a firm to redirect its strategy into different paths. Finally, in Chapter 7 we develop the main conclusions and discuss the implications and possible avenues for future research.

# 2 The Concept of Competence Building

The main purpose of this chapter is to develop an analytical framework. A review of the literature suggests that the neo-Schumpeterian approach is very relevant here in understanding the core competences of firms. Such competences enable firms to survive over long periods by continuously adapting to the changing environment and by allowing them to enter new product markets. Much of the success of Japanese electronics firms in continuously launching innovative products on the market can be attributed to their core competences. While this strategy would require not only technological skills, but organizational, manufacturing, and marketing skills, our focus is deliberately oriented towards technological competence. Given its pervasive, uncertain but strategic characteristics, technological competence is a vitally important area that firms cannot afford to ignore.

We will begin by discussing the theoretical framework which will be used in this study, namely neo-Schumpeterian theory. We will then present the main concepts involved in the notion of 'competences'. The purpose of this section is to familiarize the reader with the notion of competence. In the second part of this chapter, we present a model of competence building, outlining the factors which affect the rate and direction of the competence-building process.

## 2.1  NEO-SCHUMPETERIAN THEORY

In this section, we shall compare the neoclassical economic theory with neo-Schumpeterian theory and discuss why the latter framework has been preferable. We believe that the neoclassical economic theory is not a suitable framework for understanding the dynamics of technical change, for several reasons. The neoclassical theory assumes a static world in which firms maximize profits. Firms are characterized by a production function based on labour and capital. Firms in this theory are assumed to have perfect information and can choose from an infinite set of techniques. Technology is regarded as exogenous and treated to have many of the properties of a public good. The model

11

fails to incorporate the uncertain, trial-and-error activities of firms in developing their technical capabilities. Firms can choose from a range of production functions. The entrepreneurial activities of firms, which are diverse in nature, and therefore introduce disequilibrium in the system, are ignored in this model. The neoclassical theory focuses on demand factors and neglects the supply factors of economics.[1] Hence, while much useful and interesting work has been done within that framework, we have decided not to use the neoclassical economic theory.

In some versions of neo-Schumpeterian theory, firms are agents choosing to maximize profits at any given time from a set of alternative actions. However, there are a number of important differences: (1) Interfirm differences exist; (2) Information is not available instantaneously and freely, and each firm reacts to changing market conditions according to its decision rules; (3) Technology is not regarded as exogenous and cannot be transferred instantaneously; (4) The direction and path of technological development and learning are crucially important issues; (5) The model is dynamic. For these reasons, it offers a much more constructive framework for understanding innovation and technical change. We have therefore decided to use this framework here, since technological competence building is dynamic in character.

Modern neoclassical economic theory has largely omitted the role of the entrepreneur. Schumpeter introduced the role of entrepreneurs who innovate and capture profits, and stressed the dynamic nature of innovation, although his concepts now seem rather simplistic. Advances have been made in understanding the behaviour of firms under the framework of the evolutionary theory put forward by Nelson and Winter (1977). The core theme of the evolutionary theory rests on the dynamic process by which firms evolve through search and selection, as markets interact over time. A biological analogy in terms of reproduction, natural selection, mutation and survival of the fittest was used to describe the dynamic process of firms' evolutionary processes, 'incorporating more uncertainty, more friction, more costs of decision making and information, into the theory of the profit maximizing firm'.[2] According to the evolutionary theory, the basic assumptions of neoclassical economic theory – that firms maximize profits subject to production constraints – have been discarded in favour of organizational routines that play a role similar to genes in biological evolutionary theory.

Organizational routines are equivalent to procedures in firms; they range from procedures for training new recruits, purchasing raw materials and negotiating with financial institutions to raise funds, to policies involving R&D. Nelson and Winter propose that the routinization of activity constitutes the main way of storing organizational knowledge. Although members of a firm change from year to year as older generations of employees retire and younger ones join, evidence suggests that corporate culture persists.[3] Individuals are molded by an organization. In other words, core competences are held in organizational routines and not in individuals. Organizations remember by performing these routines. The authors distinguish between three types of routines: (1) routines that govern short-term behaviour; (2) routines that determine the period-by-period augmentation or diminution of a firm's capital stock (the selection mechanism here is analogous to natural selection); and (3) routines that relate to routine-changing processes modelled as 'searches', these being the counterpart of mutation in biological evolutionary theory.

It is unrealistic to assume, as in the neoclassical theory, that at any given time, there is a wide range of technological possibilities from which firms may choose, including alternatives that no firm has ever chosen before. Instead, in the evolutionary model any search is local, in the sense that firms choose to search in areas close to their existing techniques and technologies. The model assumes the cumulative nature of technological advance:

the output of today's searches is not merely a new technology, but also enhances knowledge and forms the basis of new building blocks to be used tomorrow. (Nelson and Winter, 1977, p.256)

The selection environment (which is a Darwinian term), refers to the environment in which some firms manage to survive while others are displaced. A very tight selection environment is one which is highly competitive. A loose selection environment may give incumbents enough 'breathing room' to develop the requisite new capabilities.[4] The direction of technological advance depends on which paths are most compelling. 'Natural trajectories', as the term is used by Nelson and Winter, are specific to a particular technology or broadly defined 'technological regime' (ibid., p.258).

The approach of Nelson and Winter was further strengthened by the addition of the concepts of 'technological paradigms' and 'technological trajectories' put forward by Dosi:

A 'technological paradigm' is an 'outlook', a set of procedures, a definition of the 'relevant' problems and of the specific knowledge related to their solution. Each technological paradigm defines its own concept of progress based on its specific technological and economic trade-offs. We will call a 'technological trajectory' the direction of advance within a technological paradigm. (Dosi, 1982)

Furthermore, Dosi suggests that, economic forces together with institutional and social factors operate as a selection device[5] (the focusing device of Rosenberg, 1976). Dosi points out that technological trajectories, or directions of technological development emerge cumulatively. In other words, the directions of technological progress within a firm or a country are related not only to the position attained, but also to the paths evolved. This concept of a trajectory is quite relevant to the competence approach. Firms develop organizational procedures which enable them to progress along those paths. We shall demonstrate later how the emerging trajectories within firms are path dependent.

## 2.2  RELEVANCE OF COMPETENCE IN STRATEGIC MANAGEMENT

### The Role of Technology in Strategic Management

Traditionally, business domains have been defined by US business schools using customer- or market-related criteria.[6] Business schools failed to realize for some time that technological changes were becoming the primary drivers of competitiveness, creating ambiguities in the definition of a business domain. However, this was the opposite of the situation in Europe, where companies had long neglected customer-oriented criteria. In recent years, there has been growing pressure on firms to give higher priority to R&D and at the same time to improve their R&D effectiveness. These pressures are associated with the accelerating rate of product innovations, shorter product life-cycles, and the growing diversification of the technological portfolio underlying the products. Several authors[7] have shown how technology diversification has been taking place at both the firm and product level. However, in many cases, firms failed to develop appropriate technology strategies.

Even boundaries between hitherto distinct sectoral segments are becoming blurred as the result of technological change.[8] For example, as Kodama (1992) has demonstrated, Asahi Kasei, a leading textile producer, is now applying its fibre technology to produce building materials. Kodama points out that high-technology companies are now facing 'invisible competitors', not knowing from which sectors competitors might come. Investments in leading-edge technologies are often necessary for survival. At the same time, new technologies are emerging which may affect many different products and processes. For management, all this implies that the traditional focus on the management of projects must be extended to include greater emphasis on the strategically more important issue of the management of technology.[9]

Technology strategy is a relation between the development of technology and the pursuit of competitive advantage in specific organizational contexts.[10] From this perspective, a business unit has three attributes; its efficiency and effectiveness, as measured by the characteristics of its products and production methods; its 'fitness', as measured by its ability to expand production and marketing capacity; and its creativity, in other words the ability to transform its underlying product and process technology. We concur with the view of Metcalfe and Boden (1990) that of these three dimensions, it is creativity which most affects the long-run survival and competitive position of the business unit.

No firm can realistically hope to invest in all leading-edge technologies, due to its financial and human resource constraints. It has to judge on the basis of what it has been best at doing, and what its competences are, to decide which technologies to invest in and which will increase its competitiveness. It also has to identify at an early stage the need for new competence building. This requires a clear focus on research, decisiveness in responding to opportunities, and leverage in using resources.

## The Emergence of the Notion of Competence

During the 1980s, a promising new approach to strategic management emerged based on the notion of core competence. According to this, there are firm-specific competences that are closely related to the firm's intangible assets, know-how and its accumulated technological resources.[11] The origins of this approach can be traced back to earlier work by Penrose (1959). Teece, Pisano and Schuen (1990) discuss the competing analytical approaches of strategic management, namely the 'competitive forces' approach of Porter, the 'strategic behaviour' approach advocated by authors such as Shapiro, the resource-based perspective, and the dynamic capabilities approach. Compared to the earlier approaches which analyzed firms primarily from the perspective of external factors such as markets and competitors, the resource-based perspective differs in the way it focuses on resources and capabilities which are internal to the firm and which are accumulated over time. Firms are seen as heterogeneous in the sense that they have differing resources, capabilities and endowments. They point out that resource endowments are 'sticky' because firms are stuck with what they have and may have to live with what they lack. 'Stickiness' arises for three reasons: (1) competence development is an extremely complex process; (2) firms lack the organizational capacity to develop new capabilities quickly[12]; (3) some assets are not tradeable.

The dynamic-capabilities approach advanced by the same authors builds upon the resource-based perspective. While the latter focuses on strategies for exploiting existing firm-specific assets, in the dynamic-capabilities approach one examines the development of new capabilities through learning and capability accumulation. It is not only the bundle of resources pointed out by Penrose (1959) that matters; it is also the mechanisms by which firms accumulate and dissipate new skills and capabilities, and the forces that limit the rate and direction of this process. Among the forces they discuss are path dependencies[13], technological opportunities, complementary assets[14] and transaction costs. We will discuss these elements in turn later in the chapter.

**A Practical Example of a Competence**

Technological competence, with its cumulative character, is increasingly important for a company to remain competitive. A simple example will help clarify what competence is. There is a small café which serves dishes such as fried egg on toast, English breakfast, omelette and chips, sausage and chips, and baked beans on toast. The customers who frequent this café are becoming more demanding, and the restauranteur faces the problem of deciding what new dish he can add to the menu. Some of his customers make suggestions including moussaka and sushi. After careful consideration, he decides against these new ventures since he has no expertise in such dishes. A friend who is a consultant asks him to write down what he has been good at doing, that is to say, his capabilities. He lists the following: (1) managing and operating a café; (2) purchasing good quality materials, specifically potatoes, raw sausages, eggs and bread; (3) quickly producing good quality products, specifically fried chips, grilled sausages, fried eggs and toast; (4) listening to customers and learning what they prefer. His friend suggests that he should think of a dish which uses these skills and provides an impetus to add to his strengths. Suddenly, a bright idea flashes across his mind, and he says 'In two weeks time, I will add hamburger and cheese omelette with chips to the menu'. Since he knows how to grill sausages, he is confident that in due course he will be able to learn how to grill hamburgers well enough to serve his customers. This dish would use his accumulated capabilities (2) and (3) and enable him to sustain competences (1), (3) and (4). Grilling hamburgers is closely related to grilling sausages. He already knows how to select good raw sausages, so it should be easy for him to purchase good quality minced meat. He will have to learn how to season minced meat and make a hamburger. Using his skills to cook omelettes, it should not be difficult for him to learn how to cook cheese omelettes. He also tells his friend that in three months time he will try to add moussaka to the menu, since he thinks he will need a few months to learn a completely new dish during his spare time. Although the skills required to make moussaka are different from those used to grill hamburgers they draw on some common techniques, such as chopping onions and whisking eggs. It also requires the ability to purchase good quality minced meat and other ingredients such as herbs. Since he is able to learn new skills, he should find it relatively easy to learn how to make the sauce, and fry aubergines. He admits that he will never be able to make sushi, since he has heard that it usually takes seven years of apprenticeship to become an experienced sushi chef in Japan. Sushi would require completely new skills for preparation as well as purchasing the raw materials which are mainly seafood.

This example illustrates how competences are path-dependent and cumulative, involve integrating a stream of skills and capabilities, and take time and effort to develop. It also highlights a firm's differential access to new technologies. Furthermore, competences involve more than technical skills, including marketing and organizational skills. Although this example was based on a simple case, one has to imagine that each skill requires a long time to develop. In the following section, we will summarize the main features of competences.

## 2.3   FEATURES OF COMPETENCES

### Different Types of Competences

There may be several types of competences, such as those related to finance, marketing and production (Figure 2.1). For example, financial competence may be defined as the ability to manage and integrate a firm's portfolio of financial assets. Marketing competence may be related to the firm's ability to integrate marketing portfolio elements, such as building a brand image and creating channels to get information on consumer demand. Production competence may concern the firm's capacity to integrate different streams of production processes. Although these competences, or 'invisible assets' are all closely related factors affecting the firm's competitiveness, we may consider them independently. For example, it is said that GEC has strong financial competence, but relatively weak marketing competence.[15] Hitachi may have strong production and technological competences but again may have relatively weak marketing competence.[16] Hitachi has been a technology-oriented firm since its founding and has emphasized R&D as being the major source of corporate growth. The fact that a marketing division was established in the headquarters only in the 1980s seems to indicate this. To give another example, two of the leading scientists who had worked on developing semiconductor lasers in the Central Research Laboratory had to go around marketing them by themselves in the US in the early 1980s.[17] Technology can affect all of these com-

*Figure 2.1*   Different types of competences

petences; perhaps the most notable example is IT (Information Technology) which is used in inventory control and financial portfolio management systems. However, we consider IT for these applications to be a technology which can be acquired relatively easily from other firms, and hence not in the category of a 'core technology' as being defined in this study.

At the centre of Figure 2.1 lies organizational competence – the firm's ability to mobilize its organization, combining people of different skills to work effectively together. The various forms of competence are coupled with organizational competence, since it is not possible to treat them completely separately. While these sub-areas of competence interact to determine the overall level of competitiveness of a firm, there is another sub-area of competence, which forms the foundation of the firm's core competence – namely, technological (or R&D) competence. In the taxonomy of Pavitt (1984), firms are classified into four categories: supplier dominated, production-intensive, specialized supplier and science-based firms.[18] Science-based firms such as concern us here are found primarily in the chemical and electronics sectors. The main sources of innovation are the R&D activities of the firms themselves. In the science-based sectors, a firm which lacks technological competence would find it difficult to be a market leader, even if it has other areas of competence, such as financial and marketing competence. If it is unable to sustain technological competences it will fail to keep up with the technological race and its products will tend to lose competitiveness. Organizations with strong and diverse in-house technological capabilities are less likely than others to be surprised or defeated by unforeseen or unmastered major innovations developed elsewhere.[19] When a firm's strength rests on its position in the market, rather than its technological bases, it is more difficult to move into new areas of specialization.[20] Nelson (1991, p.349) also underlines the importance of technological competence, pointing out that 'R&D capabilities may be the lead ones in defining the dynamic capabilities of a firm'. In a study which examined the links between corporate performance and corporate patent and patent citation data, correlations in the range of 0.6 to 0.9 were measured between increases in company profit and sales, and concentration of patents within few patent classes.[21] A firm must also have other functions including marketing, distribution and production to support the new product and process technologies arising from R&D, in other words 'complementary assets'.

A distinction needs to be drawn between 'competences' and 'capabilities'. In the study of Korea's technological capability, Westphal *et al.* give the following definition:

> Technological capability is the ability to make effective use of technological knowledge. It is the primary attribute of human and institutional capital. It inheres not in the knowledge that is possessed but in the use of that knowledge and in the proficiency of its use in production, investment, and innovation ... Technological capabilities are separable into three broad areas: production, investment, and innovation. (Westphal *et al.*, 1985, p.171)

'Production' capability is related to operating productive facilities, 'investment' to expanding capacity and establishing new productive facilities, and 'innovation' to developing new technologies.

Competence is different from capability in that it goes beyond innovation capability (which could for example come from setting up a research laboratory) to emphasize strength in particular areas, where technologies combine with processes and products as in the café example. It is what enables firms to continue and project into the future.

In countries which are in the catch-up phase, these other forms of competence such as financial or marketing may be more important than technological competence. In terms of the classification used above to describe the technical capabilities to be found in developing countries[22], it can be said that the Taiwanese have built up 'investment' capabilities. However, in order to become a market leader, during the early phase of the product cycle, technological competence is a necessary though possibly not a sufficient condition. For example, IBM's success with PCs underlined the importance of complementary assets, such as distribution networks. Thus, while the different sub-areas of competence need to be considered when evaluating the core competences of a firm, it is our intention to focus primarily on technological (R&D) competence, not only because of its importance, but because it has been a neglected area of research. It is particularly relevant in these science based industries of the kind included in this study.

**Generalizable Capability**

Recently, the concept of non-tradeable, non-imitable intangible assets has become widely recognized.[23] These intangible assets, skills and creative capabilities to create new technological capabilities are accumulated over time. As Itami and Roehl put it:

> Invisible assets are the real source of competitive power and the key factor in corporate adaptability for three reasons: they are hard to accumulate, they are capable of simultaneous multiple uses, and they are both inputs and outputs of business activities. (Itami and Roehl, 1987, p.12)

They claim that core competences of a non-tradeable non-imitable kind provide the basis for competitive advantage.

A firm should not be conceived in terms of a collection of discrete businesses. Instead, a firm should be considered as a collection of resources.[24] An advance in our understanding has been made by certain French academics.[25] They analyze the bunching strategy of technologies (*grappe technologique*), in terms of a model based on a Japanese bonsai tree (see Figure 2.2). At the lowest level lie the roots of the tree representing generic technologies. The trunk is 'the technological potential which represents the capacity to integrate generic technologies'

*Figure* 2.2    Model of bunching strategy of technologies

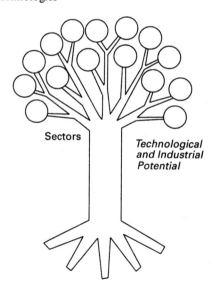

*Source*: M. Giget, Bonsai tree of the Japanese industry (1984).

and which is firm specific. Lastly, the branches and leaves of the tree correspond to the 'Sectoral *Valorisation* of technologies', enabling value to be added to the final products. In this model, the trunk represents technological competence. Technological competences of a firm depend on its capacity to evolve from new combinations.[26]

A firm's technological capability may lie upstream, where it might find applications in a variety of product applications. 'Economies of specialization' may be applied to a generalized capability.[27] We shall demonstrate later that some companies invest in building technological capabilities at the upstream end (i.e. in generic component technologies and key components). Developing an effective vertical linkage allows firms to diversify horizontally into new areas. In other words, firms which are adept at managing the linkage between products, key components and generic technologies may use these strengths to branch into often unforeseen, new areas (see Chapter 6). This is similar to the notion of an 'option value', which is a concept suggesting that investment projects do not necessarily produce additions to net present value of the firm now, but have an 'option value' for further investments at some time in the future.[28] Top management should recognize that much of a firm's R&D programme is directed toward the creation of strategic options.[29]

In a 1992 study of how Wal-Mart transformed itself to become the leader in the US discount retailing industry[30], the authors show that senior managers of Wal-Mart invested heavily in building a support infrastructure, while the senior management of its main competitor K-Mart[31], neglected to do so and followed the classic textbook approach and concentrated on making line decisions. K-Mart's managers examined their competitive advantage along a value chain and subcontracted activities that others could do better. In contrast, Wal-Mart was developing its competences by, for example, building its ground transportation fleet. Wal-Mart understood the importance of exploiting synergies across divisions and building competences. For ten years, Wal-Mart grew nearly 25% per annum, and became the highest profit retailer in the world. The authors describe the 'key component' of Wal-Mart's success as being their 'cross-docking' system, which is a logistics technique invented at Wal-Mart. In this system, goods are continuously delivered to Wal-Mart's warehouses, without ever waiting in inventory.[32] While Wal-Mart invested in the development of its 'cross-docking' system which required the acquisition of sophisticated technologies such as a satellite communications network, K-Mart subcontracted out its logistics services such as the ground transportation fleet. K-Mart gradually lost ground to Wal-Mart which was internalizing its competences. This example suggests that building competence through building capabilities in the upstream end might be generalized to other non science-based sectors.

## Definitions of the Terms Used in this Book

A firm's *Core technologies* are those underlying, enabling technologies that can affect multiple businesses because they find application in several existing business domains. Core technologies, either in isolation or in combination with other technologies, may provide the firm with the opportunity to compete in a variety of business domains.[33] Optoelectronics represents a core technology for certain firms. Core technologies are usually most critical to the survival and growth of the firm. However, while the synergy may be understood by technical specialists, they are unlikely to be recognized across all management levels.[34] Core technologies which cut across several divisions and products have to be developed and exploited through appropriate corporate and technology strategies.

We propose that at any given time, a firm may have a portfolio of core technologies $(T_1, T_2, T_3, ... Tn)$. n will vary from firm to firm. Several years later, the portfolio of core technologies may be different. Some new technologies will be added, and old ones which have become mature, discarded. By 'mature', we refer to those which have reached the plateau of an S-Curve used generally to describe technological trajectories. In other words, technological progress has reached a point where there is little scope for further innovation. n may have increased or decreased, depending on whether the firm is diversifying or narrowing its technology base. In some firms, the concept of core technologies may not be as meaningful as in others. In contrast, a company which is well known for exploiting its core technologies is Sony. A range of products which is quite distinct (for example, a video tape recorder, a walkman, and a CD player) are based on a cluster of core technologies such as integrated circuits, optoelectronics, small motor technology, and digital signal processing. Top management have a technical background and are able to manage the continuous resource allocations. In this study, we shall analyze the different ways in which firms assimilate and exploit core technologies.

There are three levels to take into account when we discuss core technologies. First, at the lowest level lie the *component generic technologies*, which are the basic *enabling technologies*. They are the technologies which are generic in nature, since they are common across a range of key components. These are combined with other technologies to produce *key components*, at the second level. These are used to develop systems and products, often in conjunction with other generic technologies. Key components are distinguished from commodity components[35] which lack differentiating potential, or the ability to develop special differentiated products. Unlike commodity components, key components add high value to end products. At the highest level lie the *systems* and *products*, which are applications dependent. For example, in the field of optoelectronics a key component such as a semiconductor laser can be used in applications from optical disks to optical communications. This is shown graphically in Figure 2.3. We intend to apply this model in the high-tech companies discussed in this study such as industrial and consumer electronics.

*Technological competence*[36] describes a firm's capacity to generate change in regard to technologies. It is a package of skills which enables a firm to respond to technological opportunities, and to assimilate them into its core capabilities. However, technological competence is more than mastery of technological capabilities. It is the ability to deploy and especially expand the range and implications of technological capabilities. It represents the relative strengths in

*Figure* 2.3 Relationship between systems, key components and component generic technologies

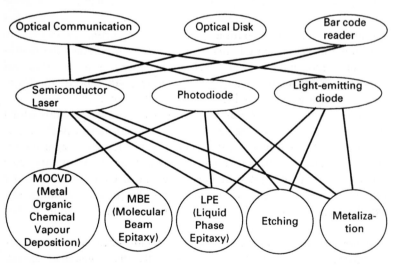

innovation capability. It requires management of a portfolio of core technologies. It is what enables firms to make use of their accumulated know-how and gradually build layers of additional knowledge to develop sophisticated key components and products. Such a strategy allows firms to enter more high value added applications and new product areas. Once competence has been built, emphasis must shift to focus on exploiting economies of scope, in order to reap economic benefits. At the same time, organizational procedures must be carried out to 'sustain' competences.

While technological competence may be one area of competence, *core competence* then involves the integration of several sub-areas of competences, thereby contributing to the firm's overall competitiveness. As we saw earlier, the knowledge generated by learning resides in *organizational routines*.[37] Routines are patterns of interactions which represent successful solutions to particular problems. Routines have a tacit dimension that often cannot be fully articulated. Hence it is the routines themselves, and the ability of the management to call upon the organization to perform them, that represent an organization's capability.

In the next section, we will put forward a model of the main factors affecting competence building.

## 2.4 FACTORS AFFECTING THE RATE AND DIRECTION OF COMPETENCE BUILDING

Using the dynamic capabilities perspective, Teece *et al.* (1990) examine the forces that focus and constrain the learning process. Among these forces are: (1) Path dependencies; (2) Technological opportunities – windows and timing of strategic actions; (3) Complementary assets and (4) Transaction costs.

Firms develop procedures which enable them to progress along trajectories. Path dependencies are not just technological in nature, and are constrained by complementary assets. In the *preparadigmatic phase* (the phase characterized by a fluid design prior to the emergence of a dominant design that becomes a widely accepted industry standard) *complementary assets* do not loom large.[38] However, in the paradigmatic phase,

as the terms of competition begin to change and prices become increasingly unimportant, access to complementary assets becomes absolutely critical. Because the core technology is easy to imitate, by assumption, commercial success swings on the terms and conditions on which the required complementary assets can be accessed. (Teece, 1986, p.196)

Contrary to what Teece claims, our evidence suggests as we shall see in Chapter 6 that core technologies are not easy to imitate. Given the time-frame of this research, we will be focusing on the preparadigmatic phase of the competence-building process, when complementary assets are not crucially important. Teece and his colleagues argue that firms are faced with brief, uncertain 'windows of opportunities' and that the development of new capabilities are constrained by transaction costs. If competence building proceeds in a cumulative, path-dependent manner, then the competence built is likely to be valuable for a specific firm. Consequently, a firm that hopes to build technological competence in a specific area *has* to invest internally. In addition, they discuss the selection environment as having an effect on firms' incentives to augment capabilities and to develop specific competences. We will outline below a model of competence building. In identifying the forces affecting the rate and direction of competence building we will draw upon the list put forward by Teece *et al.* However, since we are focusing on technological (R&D) competence, some of the factors will differ.

Competence building involves the marrying of technological opportunities with a firm's accumulated skills. I have identified a number of factors affecting this process as the following:

1. Technological opportunities;
2. Core businesses;
3. Long-term top management strategy and the evolution of the R&D organization;
4. Learning in R&D and search trajectories;
5. Management of the interlinkages between systems, key components and component generic technologies;
6. Environmental elements such as the role of government and main customers;
7. Integration.

In the next section, we will explain each factor in turn. The term 'integration' is used in three ways: the ability of the firm to integrate different technological specialisms, to combine different functional specialisms, and to exploit synergies across divisions. The term

'Knowledge base' will be used to describe the set of information inputs, knowledge and capabilities that inventors draw on when looking for innovative solutions.[39] Public knowledge is complementary to more specific and tacit forms of knowledge generated within the innovating units.

## Technological Opportunities

Several authors have advanced our understanding of the issue of technological opportunities.[40] Firms respond to technological opportunities in the belief that they should provide them with a competitive edge in transforming their products and production processes. Two types of technological opportunities are distinguished, one which is called 'intensive', and the other 'extensive'.[41] The former 'consists of improvements in an existing specification/performance relationship'. The latter is one where a particular technology 'has the possibility of being transferred into a large number of other technical systems where it can serve a variety of functions more efficiently than the existing technologies in use'.[42]

Firms may view the same technological opportunity from different angles, shaped to a large extent by their accumulated technological bases and core-business activities. Furthermore, a slanted perception of opportunities may lead to R&D resource allocations that result in a skewed technology development process.[43] Different firms may respond to the same technological opportunity in different ways. Some firms may be able to grasp opportunities as they appear, while others miss them.

Freeman and Perez (1988) argued that generic technologies can result in changes of technological systems, while others may even lead to changes in the 'techno-economic paradigm'. In an empirical study on technological discontinuities, some discontinuities have been classified as competence-enhancing and others as competence-destroying[44] :

Competence enhancing discontinuities are order-of-magnitude improvements in price/performance which build upon existing know-how within a product class ... Competence destroying discontinuities are so fundamentally different from previously dominant technologies that the skills and knowledge base required to operate the core technology shift. (Tushman and Anderson, 1988, p.93)

These two authors argued that competence-destroying advances were initiated by new entrants, while competence-enhancing advances tend to be initiated by existing firms. However, there are many instances where competence-destroying innovations are made by existing firms: for example optical disks were developed by Philips and Sony and optical fibres by Sumitomo Electric. Optical disks consist of a disk containing data in a recording layer that can be read with an optical beam. The architecture was totally different from previous stylus-based analogue disks. Not only did it reproduce clearer sound, one of its major new features was that it did not suffer from disk-to-head wear or abrasion. Pavitt's (1988a) critique of Tushman and Anderson points out that; (1) Accumulated incremental change over a long period can be difficult to distinguish from a radical change; (2) Major technical changes often grow directly out of experience in the use of earlier vintages of technology, and build cumulatively upon them; (3) The commonly held assumption (shared by Tushman and Anderson) that new vintages of technology immediately reach economically superior performance is rarely borne out in practice. A period of trial, error, learning and associated incremental change is necessary before a major new technology begins to reach its potential.

Although Tushman and Anderson classify the compact disk as a competence-destroying innovation, in fact it has drawn on many existing technologies, such as precision mechanics, surface mounting, miniaturization, digital signal-processing and small motor technologies which existed in the firms (Philips and Sony) before they began work on optical disks. In practice it is difficult to distinguish between competence-enhancing and competence-destroying innovations. However, firms cannot continue to build new technological capabilities on existing capabilities alone. At some stage, a firm will have to develop a new competence from scratch, as in the case of optical fibres which represented a completely new technology to cable and wire firms. However, even in cases which might be classified as competence-destroying, we shall show that firms are generally capitalizing on some existing core technologies; for instance, cable coating and engineering capabilities in the optical fibre example listed above.

## Core Businesses

One of the major factors affecting competence building is the firm's core businesses. During competence building, firms will search in zones that enable them to build on their existing technological bases and also on existing markets.[45] It is our understanding that competence building in firms will be centred initially in areas closely related to their core capabilities. In other words, there is dependence on the paths evolved. At a later stage of competence building, firms may use their competences built to branch into new areas.

Consider the example of path dependence as follows: Imagine someone with a scientific background, who has worked in the computer and software industries for eight years since she left University. Starting from being a programmer, she became a systems analyst, and then a project leader. She has always been working on challenging projects in an international environment. Her job takes her to Europe, America and Asia quite often. Although she finds her work exciting and rewarding, she begins to feel that her work is rather specialized and wishes she could branch out into something new. She considers changing her job and explores what are the alternatives. Although she has thought herself to be artistic, it is not too realistic for her to become a graphic designer at this point in her life, since she would have to go to an Arts College and start from scratch. She also wishes that she could be in advertising, but that too has to be ruled out, since she has no competence in that area. She realizes that she should be looking for something which builds on her accumulated strength and know-how. At the same time she would like to acquire new skills and experiences. One day she receives a phone call from a headhunter who asks her if she might be interested to work for a stock broker as an investment analyst, being in charge of the electronics and computer sectors. After careful consideration she decides to accept the offer. She would have to learn new skills, such as financial analysis. However, she would be able to use her knowhow on computer industries and software development. This example illustrates the notion of path dependence applied to an individual. In other words, the choices available to her were not infinite but limited. It depended not only on her current abilities but on the way her previous career path evolved. In the case of firms, it is similar.

Our findings conform with this approach. Even when it comes to an assimilation of a revolutionary technology, the technological trajectories developed would show features which are strongly influenced by the firm's core capabilities. This may apply to a majority of cases in 'high-tech' industries.

However, evidence suggests that there might be some exceptions, such as South Korean and Taiwanese companies. They have succeeded in developing capabilities in the high-tech sector without prior competences, through technological licensing.[46] However, in this case, they had developed 'capabilities' using the terminology of Westphal *et al.* (1985). In other words, one can acquire new competences from others, but only when one is lagging behind, and only when the basis of competition has shifted from technological leadership to price based competition.

## Top Management Strategy and R&D Organization

Competence building is an extremely complex process. Once firms start building technological competences, it may take as long as a decade or longer to achieve that task in one domain. Our evidence suggests that this is the case. It usually requires the parallel development of several sub-technological trajectories. In order to gain momentum in the competence-building process, top management in the firm must show understanding and support. In fact, competence building must take place hand in hand with a long-term corporate strategy to build technological capabilities. Firms which do not have a long-term vision or a coherent strategy might find difficulties in building competences. Although an element of chance and serendipity often plays an important role in the initial stage of competence building, the views of researchers have to be fed back into the top management strategy formulation process so that appropriate resources are allocated to build competences. In their analysis of the birth of the VTR industry, Rosenbloom and Cusumano (1987) point out that, while a Toshiba researcher did pioneering work on the VTR with a helical scanner in 1959, top management failed to understand the potential of his work and did not continue to allocate resources to the project.[47] Had the top management realized the importance of his work, Toshiba's history might have been different. Toshiba lost its early leadership to the late comers in the field (i.e. Sony, JVC and Matsushita) all of which believed in the VTR for the mass consumer market.[48] In all these firms, top management approved investing in a technology whose commercial benefits were distant in time and highly uncertain.

Considerable progress has been made in analyzing the qualities of innovative organizations.[49] They include such characteristics as: (1) the ability of the organization to accurately sense its threats and opportunities; (2) an organizational climate that fosters long-term commitment to technology; (3) willingness to take calculated risks despite the associated uncertainties; (4) spare capacity in the organization to foster the development of creative potentials;

(5) open channels of communication and contacts with outside sources of technological expertise.[50] An organization with tight control, centralized power and bureaucracy, often leads to reduced innovative behaviour.[51] Several different types of organizations exist, ranging from the classical, 'U form' organizations, to the modern 'M form'. It is said that the former type of organization is appropriate for firms operating in a stable environment where standardization and control are the norm. In contrast, the modern form is one in which participation is the mode and creativity is encouraged. This suits firms operating in an dynamic environment such as electronics.

Several authors[52] describe the tension between corporate centralization and decentralization, in other words the conflict arising from maintaining coherence and synergy on the one hand, and promoting decentralized learning on the other. Centralization is required for competence building to exploit synergies. At the same time, decentralization and rapid decision making are necessary for commercialization.

We shall show that a long-term corporate strategy is essential for building competences. Furthermore, strategy and the evolution of R&D organization are interlinked. A long-term corporate strategy will enable a gradual transformation of the R&D organization into one better suited for building competences. Firms which do not have a coherent top management strategy, instead being concerned solely with short-term return on investment, are likely to find that the R&D organization may not evolve into a form suited for building appropriate competences.

## Learning in R&D, Search Trajectories

Just as a child or an adult learns how to swim through repeated attempts, an organization can adapt itself to learn. Competence building is affected by organizational learning which involves organizational skills. It is the ability of an organization to perform tasks better through trial and error, and experimentation. Dosi and Marengo (1992) point out that

> Competences represent the problem solving features of particular sets of organizational interactions, norms and to some extent explicit strategies ... Competences are subject to learning and change through their very application to actual problem solving.

The existence of organizational learning implies that decision makers in firms cannot make optimum choices automatically, as normally assumed in the neoclassical economics. There are many varieties of learning, such as formal learning through R&D, production engineering, and other informal types described by Hobday[53] (1990). Although there has been much interest in learning, the literature on technological learning has tended to focus on learning in production and post production, rather than on R&D.[54]

For example, much work has been done on the aspects of learning in manufacturing.[55] It is generally argued that there are two types of learning in R&D[56]: First, at the basic end of R&D, the learning process involves the acquisition of knowledge concerning the laws of nature. Second, at the development end, learning involves searching for the optimum design of a product, or 'blueprinting' activity. In this study, we shall examine the competence-building process in R&D in the light of search trajectories or search paths. In other words, we shall examine the direction and scope of exploratory activities in building competence. At an early stage, competence building will be centred on the acquisition of knowledge, which may involve repetition of trial and error processes to learn about the performance characteristics of the various elements, and to deepen one's understanding of the field. We shall show that, at a later stage, the initial lengthy period of trial and error will normally begin to bear fruit, and competence building will focus on bringing innovations to the market.

The range of the search trajectories will be broad to begin with but then diminish over time as firms gradually accumulate knowledge and competence through learning. We shall demonstrate this in Chapter 6. Firms tend to search over a broader horizon in research than in technology. In the initial phase of competence building, firms are likely to explore a broad range of technological horizon as they are not sure how they would be able to gain from the technological opportunity. As firms accumulate knowledge, they would be able to direct their competence building process in areas related to their business interests.

## Linkage Between Systems, Key Components and Component Generic Technologies

We discussed earlier the importance of building competence in the upstream end. We shall show that all firms may be searching in the same common area forming the foundation of the competences being built, even though those competences will vary from firm to firm depending on their core businesses and accumulated technological bases.

For an innovation to be successful, it is often necessary to draw not just on a single core technology, but on a combination of technologies. Every technology element contains related elements and is surrounded by other related elements. Sometimes innovation in one field leads to innovation in other fields in an unpredictable way. Rosenberg (1976) gave an example showing how the type of power generation that led to greater fuel economy required high-strength, heat-resistant alloy steels. These in turn, had to await new machine-tooling methods for handling them. The notion of technological interdependence means that the various elements in a system must work together, at the same level. Having a technology or a component which stands out is of no use unless the various components are brought up to the same level.[57] One has to take into consideration the systemic aspect – the process of 'technology confluence' described by Baba (1988) and Kodama (1992).

In developing a system or a product that contains a number of components and technologies, the management has to decide which components to develop through R&D, and which to buy 'off the shelf'. There are varying degrees of importance attached to components. We distinguished 'commodity' components from key components earlier. A *key component* gives high value added to the final system, and is an integral part of the system, affecting its performance.

Baba (1988) starts from the chain-linked model[58] shown in Figure 2.4 and modifies it, focusing on the systemic aspect because 'the mode of organization of components within a system significantly influences the innovation process'.

*Figure* 2.4 Elements of the 'chain-linked model'

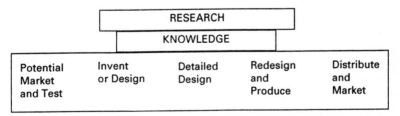

*Source*: Kline and Rosenberg, (1986); p.289. Reprinted with permission of the National Academy Press, Washington, DC; © 1986 The National Academy of Sciences.

In the chain-linked model, it is the design that triggers the innovation process. Baba's model which is shown in Figure 2.5, is based on the model used by Hitachi for formulating long-range technological strategies[59]:

> the technological requirements of a given structural element (SE) directly specify the attributes of the lower ranked ones, which are channelled in either through markets or by in-house production ... technical changes in the lower-ranked elements give rise to a series of feedback lines. These paths connect neighbouring elements and allow them to interact, or lead back directly to the higher element. (Baba, 1988)

Although the final systems and products at the higher levels of the hierarchy would vary from firm to firm, as one moves down the

*Figure* 2.5  The system organization model

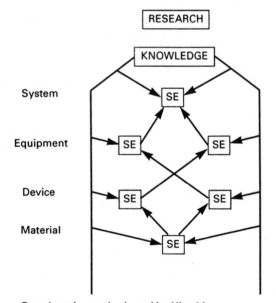

Based on the method used by Hitachi

SE = Structural Element

*Source*: Baba, (1988). Reprinted with permission;
© 1988 Y. Baba.

aggregation, we would expect to see firms actively building capabilities at the level of key components and the underlying component generic technologies, which are common to all firms. It is important to underline that building capabilities in key components and component generic technologies is crucially important for firms to become competitive at the systems level.[60]

Building an in-house capability in key components does not mean that they are made exclusively for internal use. Firms may adopt a strategy of becoming merchant makers[61] in order to make their products more competitive. If a firm's competences lie in materials, component generic technologies and key components, then the firm is likely to move gradually upward in the hierarchy to compete in systems, since they add more high value. Having a mastery over the vertical linkage of this kind illustrated in Figure 2.5 becomes an essential strategy in the competence-building process. Therefore internal communication between the teams working on components, component generic technologies and systems becomes vitally important for competence building.

Apart from organizational routines, competences may be held in other ways. Competences may be embedded in instruments, testing equipments and other hardware.[62] The accumulated knowledge about components, technologies and products may be held in the form of a database which can be accessed by employees working in the company. Competences may be held in the form of journals issued by a firm, scientific publications and patent portfolio.

During the competence-building phase, some firms are able to capitalize on economies of scope, since the merging of generic technologies provides the foundation to make a range of key components, and these in turn can be used in several applications. A combination of maintaining competitiveness in key components and generic component technologies allows firms to branch into new areas relatively easily, enabling them to capitalize on economies of scope. The range of applications of a core technology may evolve over time.[63] The skills, know-how and techniques developed for one application may be used as building blocks for developing other applications. This may take place at the level of component generic technologies, key components or systems. Furthermore, we believe that this may also involve unsuccessful attempts, as competent firms are able to learn from mistakes. The know-how, skills and techniques which have been developed for one application which failed to take off or be commercialized may nevertheless be used in successive waves of other

applications, and provide the firm with new technological opportunities. It is our belief that after the initial stage of competence building,
some firms will be able to capitalize on their know-how more than
others.

## Competence-sustaining Routines

Once competences have been built, firms must learn how to sustain
those competences otherwise erosion may occur. Competence is en-
hanced by using it. Hence the focus of learning shifts from knowledge
creation to knowledge diffusion and maintenance. This phase may be
described by 'accumulation and diffusion'.[64] Various organizational
forms can be considered in order to sustain competences. Several
studies have been carried out to examine the roles of key individuals
in successful innovations.[65] Those involved in project Sappho[66] argued
that there are four types of key individuals associated with innovation; the technological innovator, the business innovator, the chief executive and the product champion. Also important, though, are
'technological gatekeepers' whose role has been discussed by Allen
(1977) as key people who maintain long-term relationships with technologists outside the company. They have in-depth knowledge of the
scientific developments taking place in the world and are quick to get
new information which might affect the company.

While technological gatekeepers maintain links with the external
world, we would assume that, for a company to exploit core technologies successfully, there will be a mechanism to monitor them
within the firm. 'Core technology champions' as we will discuss
them later in Chapter 4 not only maintain links with technologists
outside their organization, but also coordinate activities across the
firm – for example, between different R&D laboratories, and
between development teams in the business units related to the core
technology. Since a core technology generally has multiple uses, the
core technology champion would also act as a bridge between the
divisions which might be end-users of the core technology.

## Role of Other Agents

The role of external elements of the environment such as main customers, competitors and government policies must be included as
the main factors affecting the rate and direction of the competence-
building process. A firm cannot be treated as an actor completely

isolated from its environment. Its actions are to a large extent stimulated by what happens externally. This is the key concept of the evolutionary theory with its focus on the interaction between the firm and its environment. The environment is gradually changed through the interactions. Many authors have examined the influence of market demand upon innovation.[67] Demand from the firm's main customers is likely to affect the development of technological trajectories. Rothwell (1991) points out that, in some sectors, leading-edge users can play a major role in invention and early innovation. The quality of demand from such customers will then have an influence on the search paths pursued by the company.

Some authors have discussed the role of government, and particularly of MITI, in supporting technological development in selected sectors in Japan.[68] The evidence suggests that MITI did play a central role during the 'catch up' era in the 1950s. In this study, we will briefly touch on the role of the government policies in helping the competence-building process of firms in a selected technological domain.

## 2.5 CONCLUSION

Technological competence building involves the merging of technological opportunities with a firm's core capabilities. Various factors have a significant effect on the rate and direction of competence building. Firstly, in the initial phase, competence building centres in key areas to strengthen firms' core businesses and accumulated technological bases. Secondly, the nature of competence building is extremely complex. Support and understanding by top management is essential. Top management strategy affects the R&D organization which in turn affects the competence-building process. Thirdly, competence building is affected by organizational learning. In the beginning, firms explore a broader range in research in order to learn more about the field. It takes time for firms to learn how the emerging technology could be used. As firms accumulate knowledge, the scope of competence building becomes more narrowly focused. Fourthly, there is a common area all firms actively build capabilities in, irrespective of their core businesses. Building competences in the 'upstream' end of key components and component generic technologies allows firms to be competent in systems and products. Fifthly, once competences have been built, emphasis shifts to 'sustaining' competences. Various organizational forms are necessary. Some firms actively exploit their competences built, to capitalize on economies of scope. Such a strategy allows firms to shift their trajectories to enter new areas.

This chapter has provided the theoretical foundations on which Chapters 4, 5 and 6 rest. The neo-Schumpeterian theory and the dynamic capabilities approach have been introduced. We introduced the notion of competence and its main features. Considerable attention will be given to the technological dimension of competence. Technological competence, involves the capacity to assimilate radically new technological opportunities, deploy technological capabilities and expand the range of technological capabilities. A model of competence building was put forward. Competence building is constrained mainly by two forces, which are historical, path-dependent related elements, and elements related to the present, such as organizational learning.

Chapter 4 examines the competence-building model put forward in this chapter, by drawing from the companies' case studies. In the next chapter, we shall present an overview of the evolution and techno-economic features of optoelectronics.

# 3 Optoelectronics – a Leading Core Technology

In Chapter 1, we considered the main reasons for choosing optoelectronics as a field of study. First, optoelectronics is a technological domain of growing importance and pervasiveness, yielding a broad range of applications in several key sectors of the economy. Secondly, it has been an emerging and dynamic technological domain; it is therefore suitable for studying firms' entry strategies, which are related to their previous techno-economic trajectories. Thirdly, it is also a domain which can be characterized by the three levels of (1) systems and products, (2) key components and (3) component generic technologies. It should allow us to gain an understanding of firms' strategies for maintaining effective linkages between the three levels. Finally, it is an area characterized by fierce international competition, in which Japan has done some pioneering work and in some aspects overtaken the US. We shall therefore be analyzing the competence-building strategies of European and Japanese firms in an area where Japan is at the world frontier. The main purpose of this chapter is to present an overview of the evolution of optoelectronics and its technical and economic features.

## 3.1 OVERVIEW OF THE INDUSTRY

Uses of optoelectronics can be found in many aspects of every day life. Lasers are vital equipment used in surgery; optoelectronic endoscopes allow doctors to examine parts of the body without having to perform invasive surgery; department stores and supermarkets rely heavily on bar-code readers; online database retrieval based on CD-ROMs has alleviated the work of researchers and patent analysts; and fibre-optic networks are transforming higher education by allowing teachers or students to participate in classes from remote sites. Just as transistors and semiconductor chips were major new discoveries, optoelectronics is a revolutionary technology of great importance, a core technology which must be mastered by firms in a number of high-tech industries in order to remain competitive.

**Market Growth**

Figure 3.1 plots the forecasted growth in the production value of optoelectronics in Japan, as compiled by OITDA (Optoelectronic Industry and Technology Development Association) in 1990. OITDA predicts that the production value, which increased from 90 billion yen in 1980 to 2.5 trillion yen in 1989, will reach 16.2 trillion yen by the year 2000. Table 3.1 shows the growth between 1983 and

*Figure* 3.1 Forecast of the optoelectronics market

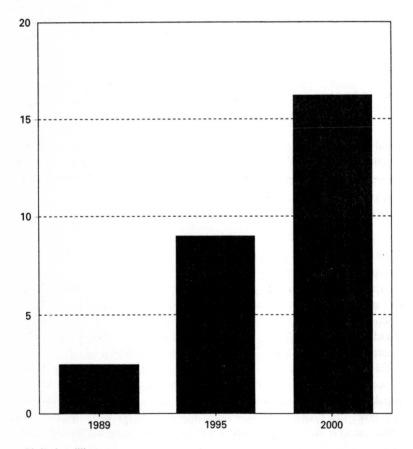

Units in trillion yen

*Source*: OITDA (1990).

*Table* 3.1   Production value of optoelectronics in Japan between 1983 and 1988

|  | 1983 | 1984 | 1985 | 1986 | 1987 | 1988 |
|---|---|---|---|---|---|---|
| **Components:** | | | | | | |
| Value | 24.2 | 27.9 | 30.1 | 33.9 | 45.9 | 47.2 |
| (% of total) | 52% | 43% | 36% | 33% | 27% | 22% |
| Annual growth | 46.1% | 15.4% | 8.4% | 12.1% | 35.5% | 2.9% |
| **Products, equipment:** | | | | | | |
| Value | 15.7 | 28.3 | 44.7 | 58.2 | 109.2 | 145.7 |
| (% of total) | 33% | 44% | 52% | 55% | 65% | 70% |
| Annual growth | 85.4% | 80.6% | 57.7% | 30.2% | 87.7% | 33.4% |
| **Applied systems:** | | | | | | |
| Value | 6.8 | 8.0 | 9.9 | 12.2 | 14.0 | 15.7 |
| (% of total) | 15% | 13% | 12% | 12% | 8% | 8% |
| Annual growth | 30.1% | 17.1% | 23.7% | 23.0% | 14.6% | 12.2% |
| **Total** | 46.7 | 64.2 | 84.8 | 104.2 | 169.1 | 208.6 |
| **Annual growth** | 66.9% | 37.6% | 32.0% | 22.9% | 62.2% | 23.4% |

Units of value: 10 billion yen.
*Source*: OITDA (1990).

1988 broken down into optoelectronic components, products and equipment and applied systems.[1]

Although the initial technological driver for optoelectronics centred on optical communications, the main impetus for mass production was led by demand for CD players.[2] Table 3.1 reflects this trend, highlighting the spectacular growth in the mid-1980s of optoelectronic products and equipment, which includes CD players.

OITDA also compiles a detailed analysis of the production value by component, product and system type. Appendix B.1 shows the breakdown in production value between 1987 and 1989. Appendix B.2 gives the breakdown as a percentage of the total value over the same period. It highlights areas of growth such as display components, CD-ROM and CD players, printers and copiers. The breakdown in 1989 indicates that 45% of the production volume is related to optical disks.

Such comprehensive data on production figures are not available for the European market. The five year average annual growth rate of the market varies across countries: in the UK it is 24%, in Germany 40%, in France 31%, in Italy 23%, and in Spain 38%.[3] The maturity of the UK market accounts for its relatively lower growth compared with other countries. In Europe, telecommunica-

*Figure* 3.2  US patents granted per year

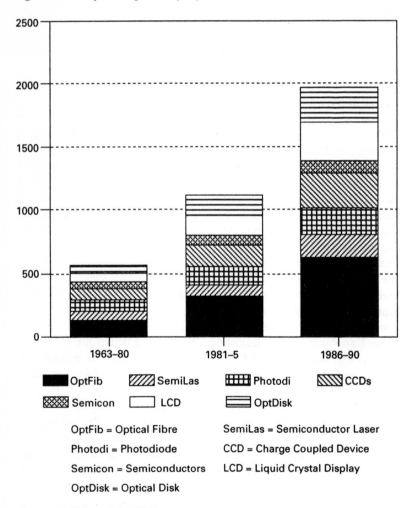

*Source*: K. Miyazaki (1991).

tions has been the main technology driver, accounting for three-quarters of the optoelectronics market in the UK. The market has grown rapidly in Germany because of Government plans to introduce optical fibres into the local network. The strength of the UK industry is primarily due to telecommunications, where BT has played a leading role. By 1986, half the trunk network was already digital. This is much higher than in other countries such as France which has developed urban networks but had not begun work on trunk networks by 1988.

Figure 3.2 plots the number of optoelectronics-related patents granted in the US over three time periods, 1963–80, 1981–5, 1986–90. It underlines the surge of US patenting in the 1980s, and clearly illustrates the rapid take-off of this technology.

## Matrix of Markets and Technologies

Table 3.2 shows the matrix of optoelectronics-related markets and technologies in the UK, and provides an example of the wide range of technologies and markets involved. However, this table may be an incomplete representation of the field since it only reflects the situation in the UK. For example, the link between lasers and consumer electronics is not identified, which might be a reflection of the weakness of the UK in consumer electronics rather than a global failure. The markets are presented in decreasing order of size. Although optoelectronics has affected other sectors such as automobiles, medicine and aerospace, in this study we decided to concentrate on the three largest sectors of application world-wide; namely, communications, information systems and industrial and consumer electronics.

In the field of communications, optical communications are used in the major trunk lines, carrying voice, TV, graphics, video, facsimile data and so on. In the field of local area networks (LANs), optical communications provide both telephony and other services. In due course, fibres will be linked to homes, offering such services as video phones and hi-definition TV broadcasting.

Optoelectronics can be found in many applications related to information systems, such as liquid crystal displays for PCs, optical storage devices and laser printers. Optical fibres are used in state-of-the-art mainframe computers in the form of optical interconnect, connecting various parts together to increase speed of operation and reduce electrical interference problems.[4]

Table 3.2  Matrix of optoelectronic-related markets and technologies

| Technol. / Markets | Material | TransSw | CompoC | Op.Info.Pr | Op.Stor | Displays | Imaging | Sensors | Lasers |
|---|---|---|---|---|---|---|---|---|---|
| Comm | *** | *** | *** | *** | * | ** | * | | ** |
| Info Sys | ** | ** | ** | *** | *** | *** | ** | | * |
| Consumer | ** | | | | *** | *** | ** | * | |
| Military | *** | *** | *** | *** | *** | *** | *** | *** | *** |
| Automo. | ** | * | ** | | * | *** | | ** | |
| Aerospace | ** | * | ** | * | * | *** | | ** | |
| Medical | ** | | * | | *** | ** | ** | *** | *** |
| Mat. Proc | ** | | | | | | | *** | *** |
| Proc. Cont | ** | ** | ** | | * | ** | ** | ** | * |
| Safety | ** | ** | ** | | * | * | *** | ** | |
| Energy | *** | | | | | | | | * |

TranSw =Transmission and Switching systems; CompoC=Components for communications systems; Op. Info. Pr= Optical information processing Op.Stor= Optical storage of information; Comm=Communications; Info Sys=Information systems; Automo.=Automotives; Mat. Proc=Materials processing; Proc. Cont=Process control.
Importance of technology to market *** High. ** Medium. * Low.

*Source:* Adapted from ACOST (1988). Crown copyright. Reproduced with the permission of the Controller of Her Majesty's Stationery Office.

In the field of consumer electronics optoelectronics is used in products such as flat-screen displays based on liquid crystal displays (LCDS) for TV sets, pocket size TVs, projection TVs, compact disks and videodisks. CD-ROMs have also become widely available as an information-retrieval device. Recently, Kodak has launched a new photography system based on optical storage techniques.

In the military field, fibre optics has increasingly replaced metal cabling for military communications. Optoelectronics is incorporated in imaging sensors, thermal imaging and range finding.

## 3.2 INTERNATIONAL COMPARISONS

Table 3.3 provides the relative standing of US, Japan and Europe in emerging technologies, compiled by the Department of Commerce in the USA.

As Table 3.3 shows, optoelectronics is a field where Japan and the US are level in terms of R&D, but one in which Japan may be leading in terms of product innovation. We shall therefore be examining the competence-building strategies of firms in a field where Japan is at the world frontier. Hence we are examining a large share of the world total stock of technological knowledge in the field that Japan has caught up with or in some cases overtaken the US. This is shown by Table 3.4, which shows the number of patents granted in the US over three periods, 1963–80, 1981–5 and 1986–90 – broken down by inventor country, for each of the key areas of optoelectronics.

Table 3.5a lists the percentage of total patenting in the US by the two countries, US and Japan, for the whole period under consideration, while Table 3.5b lists data for the most recent period only (1986–90). Within these tables, a number of significant patterns and trends can be detected. First of all, the US and Japan account for approximately 70–80% of the total patenting in all fields. The rest is shared by European countries such as Germany, France, the UK and the Netherlands. In other words, patenting is highly concentrated in this field.

One of the most important trends in these tables is the rapid increase in Japanese patenting in the US, in all the fields shown. A comparison of shares in recent years to shares of total patenting for all countries suggests that Japan has gained in relation to the US in every field. In some key areas, such as optical fibres, semiconductors and epitaxy, the US continues to lead, but the gap has been

*Table* 3.3 Relative standing in emerging technologies: US versus Japan and the EC

| | *Japan* | | *European Community* | |
|---|---|---|---|---|
| | *R&D* | *Product* | *R&D* | *Product* |
| Advanced Materials | OL | –L | +H | OL |
| Advanced Semiconductors | OH | –L | +H | OH |
| AI | +H | +H | +G | +H |
| Biotechnology | +L | +L | +G | +H |
| Digital Imaging | OL | –L | OL | –L |
| CIM | +H | OH | +L | –L |
| HD Data Storage | OH | –L | +H | OH |
| High Perf. Comp. | +H | +L | +G | +G |
| Medical Device | +H | +L | +H | +G |
| Optoelectronics | OH | –L | OH | +H |
| Sensor Technology | +L | OH | +H | OH |
| Superconductivity | OL | OL | OH | OH |

United States versus Japan and the EC:

| Current status: | Trend: |
|---|---|
| + = US Ahead | G = US Gaining |
| O = US Even | H = US Holding |
| – = US Behind | L = US Losing |

AI = Artificial Intelligence; CIM = Flexible Computer Integrated Manufacturing; HD Data Storage = High-density Data Storage; High Perf. Comp. = High Performance Computing; Product = Product Innovation.

*Source: Emerging Technologies, A Survey of Technical and Economic Opportunities* (1990), Dept of Commerce, Washington, D. C.

narrowing. In other areas, such as semiconductor lasers and photodiodes, Japan has overtaken the US in the most recent (1986–9) period. In liquid crystal displays and optical disks, Japan caught up with the US in 1981–5, and has since then become the world leader. In both these fields, Japan holds over 50% of total US patents for 1986–90.

*Table* 3.4    US patenting in optoelectronics: international comparisons

**Semiconductors Epitaxy**

|  | 1963–80 | 1981–5 | 1986–90 | Total |
|---|---|---|---|---|
| USA | 588 | 193 | 260 | 1041 |
| Japan | 139 | 86 | 158 | 383 |
| W. Germany | 54 | 15 | 13 | 82 |
| France | 38 | 12 | 23 | 73 |
| UK | 26 | 19 | 12 | 57 |
| Netherlands | 20 | 3 | 2 | 25 |

**Liquid Crystal Displays**

|  | 1963–80 | 1981–5 | 1986–90 | Total |
|---|---|---|---|---|
| USA | 702 | 240 | 398 | 1340 |
| Japan | 297 | 299 | 797 | 1393 |
| W. Germany | 68 | 91 | 123 | 282 |
| UK | 46 | 49 | 78 | 173 |
| Switzerland | 81 | 45 | 38 | 164 |
| France | 56 | 35 | 55 | 146 |

**Optical Disks**

|  | 1963–80 | 1981–5 | 1986–90 | Total |
|---|---|---|---|---|
| USA | 433 | 312 | 342 | 1087 |
| Japan | 116 | 329 | 877 | 1322 |
| Netherlands | 84 | 72 | 69 | 225 |
| W. Germany | 48 | 30 | 45 | 123 |
| France | 47 | 39 | 34 | 120 |
| UK | 14 | 5 | 18 | 37 |

**Optical Fibres**

|  | 1963–80 | 1981–5 | 1986–90 | Total |
|---|---|---|---|---|
| USA | 1793 | 899 | 1512 | 4204 |
| Japan | 243 | 228 | 689 | 1160 |
| W. Germany | 188 | 159 | 312 | 659 |
| UK | 171 | 93 | 197 | 461 |
| France | 130 | 113 | 156 | 399 |
| Canada | 30 | 41 | 78 | 149 |

*Table* 3.4—*continued*

Semiconductor Lasers

|  | 1963–80 | 1981–5 | 1986–90 | Total |
|---|---|---|---|---|
| USA | 606 | 185 | 331 | 1122 |
| Japan | 156 | 131 | 426 | 713 |
| W. Germany | 33 | 13 | 53 | 99 |
| France | 38 | 10 | 26 | 74 |
| Netherlands | 13 | 11 | 16 | 40 |
| Canada | 10 | 10 | 9 | 29 |

Photodiodes

|  | 1963–80 | 1981–5 | 1986–90 | Total |
|---|---|---|---|---|
| USA | 1171 | 420 | 424 | 2015 |
| Japan | 229 | 186 | 413 | 828 |
| W. Germany | 72 | 42 | 56 | 170 |
| France | 71 | 25 | 47 | 143 |
| Netherlands | 38 | 17 | 17 | 72 |
| Canada | 19 | 12 | 16 | 47 |

CCDs, Image Sensors

|  | 1963–80 | 1981–5 | 1986–90 | Total |
|---|---|---|---|---|
| USA | 1101 | 461 | 631 | 2193 |
| Japan | 117 | 193 | 407 | 717 |
| W. Germany | 99 | 70 | 72 | 241 |
| UK | 52 | 31 | 67 | 150 |
| Netherlands | 29 | 16 | 50 | 95 |
| Switzerland | 11 | 28 | 14 | 53 |

*Source*: K. Miyazaki (1991).

Table 3.5a    *Percentage of US patents by country (1963–90)*

|        | OptFi | SemiLa | Photodi | CCD  | Semico | LCD  | OptDi |
|--------|-------|--------|---------|------|--------|------|-------|
| US     | 56.7  | 50.8   | 58.2    | 58.2 | 61.5   | 36.7 | 36.6  |
| Japan  | 15.7  | 32.3   | 23.9    | 19.0 | 22.6   | 38.2 | 44.5  |

Table 3.5b    *Percentage of US patents by country (1986–90)*

|        | OptFi | SemiLa | Photodi | CCD  | Semico | LCD  | OptDi |
|--------|-------|--------|---------|------|--------|------|-------|
| US     | 48.1  | 36.9   | 41.4    | 45.9 | 54.9   | 25.6 | 24.3  |
| Japan  | 21.9  | 47.5   | 40.3    | 29.6 | 33.3   | 51.2 | 62.3  |

*Source* : K. Miyazaki (1991).
OptFi=Optical Fibres; SemiLa=Semiconductor Lasers;
Photodi=Photodiodes; CCD=Charge Coupled Device;
Semico=Semiconductors; LCD=Liquid Crystal Displays; OptDi=Optical
Disks.

## 3.3   THE TECHNOLOGIES OF OPTOELECTRONICS

**Technological 'Fusion' and Core Competence**

Optoelectronics involves the manipulation of photons[5] and their interaction with electrons. It uses light to perform a variety of functions, such as transmitting, storing and processing information. Electrons and photons have distinctive properties which give rise to various advantages and disadvantages for different functions. For example, photons travelling at the speed of light are scarcely attenuated by insulators, and so are ideal for long-distance transmission of information. Electrons, on the other hand, interact with each other and with most materials, and can be finely controlled in ways suitable for information processing.

Let us first discuss whether optoelectronics – the interaction between photons and electrons (optics and electronics) – can be considered in the same way as mechatronics[6]. According to Kodama (1986a), mechatronics is an example of technological fusion, which has been made possible by a concerted effort of the several different industries involved. The mechatronics revolution was brought about through the cooperation of three industries, in addition to the machine tool industry itself – that is, precision instruments, communications and electronics. However, optoelectronics is a different case from mechatronics. Kodama's notion of mechatronics stresses the importance of merging technologies through pooling resources across industrial sectors. In contrast, optoelectronics exemplifies the merging of optical and electronic properties at the atomic level. It follows that optoelectronics should not be considered in the same way as mechatronics. This is because optoelectronics is a case of 'scientific fusion' and should be regarded differently from Kodama's case of 'production fusion'. However, technology fusion does take place in the case of optoelectronic applied systems, such as optical disks, with the merging of several technologies, including disk media, laser diodes, precision mechanics and digital signal processing. The key components of an optical disk, which are the laser diode and the pick-up, are based on the principle of optoelectronics pointed out above.

Great progress has been made in optoelectronics over the last thirty years and it is becoming a core technology, underpinning advances made in communications, information systems, consumer electronics and other sectors. In the development of electronic equipment for industrial users and consumers, such as personal computers, word processors and TVs, technological leadership in peripherals is becoming crucial, partly because they are part of the human interface, but also because there has been a trend to embed more functionality in the peripherals.[7] For example, the performance of personal computers and TVs is determined to a large extent by the display mechanism, while a mainframe computer's performance depends more on its capacity to store and retrieve large amounts of data. In what follows, we shall examine the relationship between the underlying key components and component generic technologies that are required to develop the core products and systems affected by optoelectronics in the chosen sectors. We have focused particularly on component generic technologies, rather than system-related generic technologies, which might be different. For example, controlling the network as a whole involves system reliability technologies such as software.[8]

In the next section, we shall discuss the main systems and products, giving an account of their evolution, functions and the main advantages and disadvantages compared with other more conventional competing technologies. We shall then outline the evolution and function of the main key components and component generic technologies.

**Optical Communications**

In 1966, Kao and Hockham of STL (Standard Telephones and Cables Laboratory) proposed the revolutionary concept that optical communications could become a possibility. They showed theoretically that, if the optical fibre transmission loss[9] could be reduced to below a certain threshold level, optical communication could be achieved. At that time, typical fibre losses were of the order of 1000 db/km[10], two orders of magnitude higher than the threshold level of 10 db/km required. Their announcement triggered a race between the national PTTs (Post, Telephone and Telegraph companies) in Europe, US and Japan to develop optical communications technologies.

Optical communications became a reality in 1970, following advances made in two parallel fields. First of all, semiconductor lasers oscillating at room temperature were discovered by Panish and Hayashi[11] working at Bell Labs. Although laser action in semiconductors had been known since 1962, it was only in 1970 that continuous operation at room temperature became possible. In the same year, the possibility of low-loss optical fibres was demonstrated by Corning. These two parallel developments led to the demonstration of communication systems that could transmit information at a rate of over two billion bits per second over 130 km, with an error rate of one bit per billion.[12] A single fibre could thus carry over 10 000 telephone conversations over a distance of 100 km at a lower cost than any other technologies. Optical fibre networks were first applied in industries such as steel, electric power and electric machinery. Joint collaboration between the electric power companies and the electronics companies such as NEC began in the mid-1970s. The first trial of long-distance, public optical communications took place in the US in 1976 using Corning fibres. In the following year, Britain became the first country to carry public telephone traffic through optical fibres.[13]

Although in the initial stage of development, optical communications centred around the 0.8 μm wavelength band, it was shown in 1975–6 that light in the 1.3 μm wavelength region was most transparent[14] in silica fibre, hence more suitable for long distance optical communications.[15] This finding was reported in two papers, one by researchers at Southampton University in the UK and the other by NTT (Nippon Telephone and Telegraph Company).[16] While GaAlAs/GaAs[17] lasers were suitable for the 0.8 μm region, the materials were based on InGaAsP/InP[18] for operating in the 1 μm region. We will see later that most Japanese firms engaged in optical communications switched to the new materials shortly after in 1977–8. At first, it was found that the dispersion which caused transmission losses, became almost zero at 1.3 μm wavelength, and then later that it was even lower at 1.55 μm. Thus, optical communication switched from short to long wavelength.

In Japan, NTT completed an optical transmission field trial in 1977. At that time, Japan was ahead of the rest of the world in conducting field trials of long wavelength optical communications; while other countries began to catch up only at the beginning of the 1980s.[19] By 1985, optical fibre transmission systems had been widely introduced into the main trunk lines with over 4000 km of optical cables installed. Optical communication systems have been implemented in a variety of applications, including trunk lines, optical Local Area Networks, and replacing conventional wire harnesses[20] in automobiles.

Several basic building blocks are required to achieve optical communications. A light-emitting element converts an electric signal to light; a light-transmission element transmits light through a medium; a light-receiving element receives light and converts it back to electricity.

The technological trajectory of optical communications has centred on: (1) increasing transmission capacity, and (2) increasing transmission distance. The individual technological trajectories of various key components that make up optical communications have been strongly influenced by these goals.

By examining the technological trajectories, we find that optical communications can be divided into three phases – the early, mid- and late phases. In the early phase, a gallium arsenide based LED[21] was used as a light source at 0.85 μm wavelength. The distance between repeaters was short (about 10 km). In the mid-phase, indium gallium arsenide phosphide-based semiconductor lasers were used as the light source, operating at a wavelength of 1.3 μm. The distance between repeaters was extended to 30–300 km. In the late phase, the wavelength has changed to 1.55 μm, and the aim has been to achieve an 'all optical' communications system by using optical fibre amplifiers to boost the optical signals travelling down

*Table* 3.6   The three phases of optical communications

|                   | Early(1970s-) | Mid-(1982-)         | Late(1988-)         |
|-------------------|---------------|---------------------|---------------------|
| Light Source      | LED           | Semiconductor laser | Semiconductor laser |
| Wave Length ($\mu$m) | 0.85       | 1.3, 1.55           | 1.55                |
| Distance (km)     | 10            | 30–300              | 100–300             |
| Other Features    | Multi-mode    | Single-mode         | Optical Amplifier   |

the optical fibres directly without first converting them back to electric signals.

Optical communication has three main advantages over conventional forms of communications based on metallic cables. First, they provide larger bandwidth and suffer lower losses (they can transmit up to 100–1000 times the amount of information carried by conventional metallic cables). Secondly, optical fibres weigh less than one hundredth of metallic cables, and their small volume leads to lower construction costs. Thirdly, optical signals are immune from electrical interference. Such immunity is important for military applications since it permits cables of highly sensitive electronics to be located near electrical power system wiring on land, at sea and in aircraft. On the other hand, optical fibre communication has two principal disadvantages. First, it is much more difficult to connect optical fibres than copper cables. Secondly, a minute bend in optical fibres may lead to transmission losses, so they need to be well protected against stress as well as temperature variation.

**Display Devices**

Although liquid crystal materials were discovered almost 100 years ago, interest in their use as display devices developed only in the late 1960s. The first liquid crystal devices were invented in Switzerland, followed by the first prototype display device developed by RCA in the US in 1968. However, the product was technologically premature and the company gave up commercializing it.[22] In the UK, the MoD (Ministry of Defence) set up a Working Party in 1967 to study an alternative to the CRT (cathode ray tubes).[23] Liquid crystals emerged as a possibility, and a collaborative programme was initiated in 1970 between Hull University and RSRE (Royal Signal and

Radar Establishment) to develop liquid crystal displays. This project succeeded in developing the first generation of liquid crystal materials and also components suitable for watches and calculators. The materials were patented in 1972 and passed to BDH Chemicals Ltd.[24] Cyril Hilsum, who had been the chief scientist at RSRE, later joined GEC in 1982 to start a research programme within the company.[25] Although much of the pioneering work on liquid crystal displays was done in the UK, large-scale manufacturing in the UK has been limited. It is Japanese firms which have been much more successful in the commercialization and mass manufacture of LCDs. As will be discussed in Chapter 4, the British firms may have been more risk averse compared to the Japanese companies when they had to decide whether to enter the LCD market for the mass market.

Liquid crystal displays are formed by filling a cell made by two glass plates with liquid crystal material. DSM (Direct Scattering Mode) was the first LCD display mode used in calculators. Although these displays required relatively high driving voltages and therefore had high power consumption, they were physically attractive and were used especially in watches. In the early phase, it was merely necessary to display numbers. As more sophisticated applications developed, display of alphanumeric characters and graphic symbols was required and multiplexing[26] became important. DSM was gradually replaced by twisted nematic (TN) displays which dominated over the next fifteen years because of the low driving voltage which they required.[27] TN LCDs eventually reached a resolution limit of about $10^4$ pixels.[28] Supertwisted nematic (STN) displays then gradually displaced TN displays in the mid-1980s. They are capable of achieving improved contrast, a wider viewing angle, faster response and clear colours. Appendix C shows how the usage and dimensions of LCDs have changed over time.

To overcome the above problems such as faster response and clear visibility and for making larger portable LCDs, the manufacturers have developed active-matrix LCDs, in which each pixel (or picture element) is controlled by a switching element such as a TFT (thin film transistor). Thus, for a 640 × 480 pixel display, there would be 307 200 TFTs, one located at each pixel. Two materials are mainly used in Japan for the active layer in TFTs – amorphous silicon and poly-silicon. Active matrix displays are a good candidate for high definition TVs. Increasing pixel density in a TFT LCD leads to decreasing display brightness, one of the problems to be overcome in developing large LCDs. Another problem to be overcome is the low fabrication yield and high production cost as the size of LCD increases.

LCDs can be regarded as a form of technology fusion. Liquid crystal displays consist of a thin film of liquid crystal contained between parallel, electrically conducting plates and placed between parallel or crossed polarizers. If an electric field is applied across the film, the molecules realign and the plane of polarized light transmitted through the film is rotated to produce bright or dark states in the analyzer.

In the early applications, there were four main research objectives, namely the materials, temperature range, contrast ratio and increasing life. However, as the applications grew more and more sophisticated the range of issues became more complex. LCD requires parallel development of several technologies, including materials, back light to illuminate the display, microfabrication processing, electrical connection between the LCD panel leads and driver circuits, IC drivers, polarizers, thin film growth such as plasma CVD (Chemical Vapour Deposition), lithography, and 'ultra clean room' technologies.[29] These technologies have been developed over a long period, with many incremental improvements. As LCDs became more sophisticated, the range of generic technologies drawn upon became broader and technically more advanced. A number of technologies have become mature (such as driving circuit, polarizer, and TAB (Tape Automated Bonding), while others are still in the development phase, (such as materials, back light, Chip on Glass and TFTs).[30]

Compared to plasma and electroluminescent displays which are emissive displays, LCDs are of a subtractive type which absorb incident light. The technology of emissive displays is difficult and may not greatly increase its small market in the long run.[31] Electroluminescent displays have been used as a display device for high performance engineering workstations although they have the drawback of a limited colour range.

TN LCDs have two main limitations – poor visibility and narrow viewing angle. However, they retain some advantages, including stability over a wide range of temperatures and ease of production. LCDs have become widely used as a display device because of their thinness, light weight and low power consumption. In the case of the CRT, only three circuits are needed – the brightness modulation circuit, and the vertical and horizontal deflection circuits – irrespective of the display size. The price of a CRT does not vary much with the display area. While the CRT is cheap, focusing tends to be blurred, suffers from flickering images, and it is also bulky. The features and defects of the flat-panel displays such as LCDs are quite different. Prices vary with size: the bigger the display size, the higher the price, and the lower the production yield. For example in making a display with 640 by 400 pixels, 640 × 400 × 3 (one for each colour) TFTs have to be attached to the panel without a single defect. Focusing is sharp, and the display does not suffer from flickering images. It benefits from compactness and slim size. Apart from display devices for PCs and TVs, other important applications for instrumentation displays are in automotive and avionic systems where the display devices need to meet more demanding environmental conditions.

**Optical Disks**

Using the properties of light to store data has been envisaged by many researchers since the 19th century. However, it took almost a hundred years for optical storage to become a reality. In the 1960s, both Toshiba and Hitachi worked on the application of holography, using lasers to form optical memory storage devices, although these were not commercialized. Both companies began work on optical disks in the 1970s, Hitachi in 1974 and Toshiba four years later.

Philips launched a research project on videodisks in the late 1960s and were able to develop a prototype in 1972, using gas lasers. Sony entered the field of optical disks in 1974. During the next five years, eight to ten people worked on a project involving laser disks, developing various technologies such as lithography, mechatronics, sputtering, optical disk media and digital signal processing. In 1978–9, Sony and Philips agreed to collaborate to work on a compact disk player and Sony succeeded in launching it in 1982. Videodisks also became a popular audio product. Subsequently, successive generations of compact disk players have been put on the market. After audio disk technologies, companies have moved on to develop optical storage technologies, such as CD-ROM, which was commercialized in 1986, and CD-I (CD Interactive) which was launched in 1987.

Optical disk technology has been suited for use as a low-cost mass storage device. It is capable of handling several types of data, including audio (compact disk), video image (laser disk), text (CD-ROM), graphic and text (CD-I). An optical disk consists of a disk containing information in the form of marks in a recording layer that can be read with an optical beam. The basic components of optical storage consist of (1) a clearly modulatable light source such as a laser diode, (2) focusing optics and a tracking system, (3) data encoding and decoding, (4) error detection and correction, and (5) information detection (e.g. phase change, intensity change, polarization change) systems.

The WORM (Write Once, Read Many times) optical media are user-recordable. In a WORM system, a focused laser beam causes pits to be formed on the surface of the disk. These pits can be read with a laser pickup. CD-ROMs, in contrast, are not user recordable, but are 'read only' disks. The data are provided by a data provider and replicated on the disks. They are used for information retrieval and electronic publishing. A more recent application of optical disk technology is found in erasable disks. The data on the disk can be erased and rewritten, using magneto-optics technology. A laser beam heats the spots on a disc and a magnetic field is applied changing the direction of the magnetization at these spots.

Optical disk storage offers the advantages of (1) lower cost per bit of storage (the cost per bit of storage is one quarter of that of floppy disks), (2) higher storage capacity of the order of giga bytes[32], (3) long life, (4) no disk-to-head wear or abrasion since there is no direct physical contact as in the case of records, (5) compactness and ease of use, and (6) ability to store video, data, and audio on the same disk. On the other hand, optical storage may have the following disadvantages (1) relatively slow data transfer rate (the figures for CD-ROM and a 40 MB hard disk are respectively, 150 and 625 kbytes/sec); and (2) relatively slow data access time.[33]

## Key Components

### (a) Light-emitting diodes

Of the key components, light-emitting diodes (LED) were among the first to be commercialized. Monsanto of the US marketed the first display-type LED in 1968. Between 1968 and 1980, a number of technical improvements were made for the mass production of crystals while the method of liquid epitaxial[34] growth was developed. A variety of LEDs of different colours, including green, red, and yellow were developed. A growing demand for display-type LEDs led to high annual growth rate of 130–150% in production volume in Japan between 1976 and 1979.[35]

There are two types of LEDs, one used mainly as an indicator display, and the other in communications. In the former, demand exists for use in VCRs, audio units and OA (Office Automation) equipment. In the early days of development, LEDs were used to display 'points', and then they were developed into 'lines'. Later, LEDs have come to be developed into XY arrays[36]. More recently, demand has expanded to LED printers, facsimiles, automobile display units and traffic signal units.

In the second type, LEDs are used for short-distance communications. In these type of LEDs, the structure is simple compared to that of a laser. Infrared-emitting diodes are used in the range up to 200 Mbit/s and 2 km span. While infrared diodes emitting in the 0.8 $\mu$m range are based on GaAlAs[37], those emitting in the 1.3 $\mu$m range are based on InGaAsP[38]. In many cases a glass or a sapphire sphere on top of the chip is used to focus the emissions onto the front face of the fibre, thereby bridging the gap from chip to fibre. Infrared LEDs are also used as remote control devices, for example for TV sets.

*(b) Semiconductor lasers*

Since the discovery of room temperature laser oscillations in 1970, the semiconductor laser has become the key component of optoelectronics in the field of communications and industrial and consumer electronics. In the early days of research (up to 1976) improving life and finding out the causes of defects were the most pressing tasks. This was followed by other objectives such as obtaining single-mode oscillation[39] and improving quantum efficiency. In 1978, research on semiconductor lasers shifted to work on a different material based on indium gallium arsenide phosphide in the 1.3 $\mu$m region.

In Japan, some 10 million semiconductor lasers were manufactured in 1987 for use in optical communications equipment and CD players.[40] This presents a sharp contrast to the fact that the output of solid lasers and gas lasers is limited to several thousand units. Appendix B.1 also shows that the market size of semiconductor lasers in Japan was three times the size of gas lasers and ten times the size of solid state lasers in 1989. This difference is because semiconductor lasers have the advantages of small size, light weight, low cost and low power consumption, features which are lacking in solid or gas lasers.[41] In this study, we have limited the analysis to semiconductor lasers, since they are much more important in communications and industrial and consumer electronics than other types of lasers. Furthermore, compared to other components such as LEDs which have become mature products in the sense that products have become fairly standardized, semiconductor lasers continue to be in a state of development.

There are two categories of semiconductor laser – long wavelength and short wavelength – depending on the wavelength of light emitted. Lasers consist of a medium in which light amplification takes place between a pair of mirrors forming a resonator. These lasers are made from heterostructures[42] in which a thin active region is formed between passive cladding regions. Elements from the family of III-V semiconductors[43] such as GaAs (Gallium Arsenide) or AlGaAs (Aluminium Gallium Arsenide) are commonly used to form these thin layers.

Semiconductor lasers in the long wavelength band (1.3 μm, 1.5 μm) are used for long-distance trunk-line communications. Materials such as InGaAsP (Indium Gallium Arsenide Phosphide) are used to form lasers in the 1.3–1.5 μm region. In 1988, the market size for semiconductor lasers in the 1.3 μm band in Japan was 24 000 units, while demand for lasers in the 1.5 μm band was only 1300 units since the latter type was still in the research and development stage.[44] The unit cost of semiconductor lasers (of the DFB type) for optical communications ranged between 100 000–500 000 yen in 1990.[45] The average price of lasers used in the 1.3 μm band dropped from 100 000 yen in 1980 to 20 000–30 000 yen in 1990.[46]

Semiconductor lasers in the short wavelength band (0.6 μm – 0.8 μm, visible region) are used as a light-emitting source for optical disks (such as compact disks and CD-ROMs) and laser printers. A wavelength of 0.78 μm is used as a light source for the compact disk player, which was first launched in 1982. CD players have been becoming smaller every year. It is therefore important that optical pickups become smaller and thinner. Prices of semiconductor lasers used in CDs fell from 3000 yen to 300 yen over the same period, signalling that they have become commodity components. It is possible to increase the recording density of optical disks by shortening the wavelength. In optical disk applications, noise suppression and reduction of the wavelength have been the most important goals, while for undersea communications, increasing the product/service life and improving reliability have been of paramount importance, since it is very costly to repair an undersea optical fibre link once the system is operational. Increasing the output power has also been important in optical memory disk applications. Applications of high-power lasers

(50 mW) can be found in optical information processing including optical disks and laser printers. Such lasers are normally based on AlGaAs material.

### (c) Light-receiving key components – photodiodes

In the early development, when optical communication centred on the 0.8 μm wavelength region, research focused on development of silicon PIN[47] photodiodes or avalanche photodiodes (APDs). In the latter half of the 1970s, research shifted to thin film epitaxial growth techniques of photodiodes and Ge detectors in the 1.3 μm region. The technological development of the materials and structures involved is closely related to that of lasers.

Light which has been transmitted at one end of an optical fibre has to be detected at the other end. Photodiodes receive light-pulse signals and convert them back to electric signals. For short distances, Silicon (Si) based photodiodes are used, while in the long wavelength range, Ge (germanium) or GaInAs (Gallium Indium Arsenide) semiconductor, avalanche photodiodes are employed. In the 1.5 μm band, light-receiving components such as PIN Photodiodes (PIN-PD) are important. PIN-Photodiodes have several advantages such as their low-voltage operation, and high-diode sensitivity.

### (d) Transmission Elements – Optical Fibres

The earliest optical fibres developed were of multi-mode[48] type. In 1970, silica-based multi-mode optical fibres with a transmission loss of 20 db/km at 0.8 μm wavelength were developed. The development of single-mode optical fibres began in the late 1970s and these were commercialized in the early 1980s. Mono-mode (single-mode) fibres are well suited for high data rate, long-distance communications since they do not suffer from multi-mode dispersion. Over the last 10 years, the fabrication technology for silica glass optical fibres has improved and a loss figure of 0.2 db/km at 1.5 μm has now been achieved. In the UK, one individual in the British Post Office, Dr O'Hara decided early on that single-mode fibres were the most promising and, against strong opposition both from within the Post Office and from the suppliers such as STC, forced that path to be pursued. He was eventually proved right and the eminent UK position in single-mode fibres is due largely to his perseverance.[49] The technological trajectory has been driven by the search for decreasing dispersion and losses. There has been a gradual improvement in the quality of materials, so that losses due to impurities have been minimized almost as far as is physically possible. Optical fibre technology based on silica has now become well developed and can therefore be considered as a mature technology.

Three types of fibres are mainly produced: multi-mode graded-index, multi-mode step-index and single-mode fibres. Several types of optical fibre production method exist: the OVPO (Outside Vapour-Phase Oxidization) process developed in 1972 by Corning, the MCVD (Modified Chemical Vapour Deposition) method developed by Bell Labs in 1974 and the VAD (Vapour-phase Axial Deposition) developed in Japan by NTT and several leading cable makers[50] in 1975. Compared to the previous methods, in the development of VAD, special attention was paid to the following objectives: (1) realizing a continuous fabrication process; (2) mass production feasibility; (3) high fibre performance; and (4) process simplicity[51]. Thus the VAD method made it possible to produce high silica fibre preforms[52] continuously and to make larger preforms improving the yield. The cost of fibre in cable form dropped by a factor of over ten since 1980, thanks largely to improvements in manufacturing techniques and increased volume of production.

Various types of optical fibres exist, such as quartz optical fibres, multi-component glass optical fibres, plastic optical fibres, infrared fibres, image fibres and OPGW fibre which was invented by Sumitomo Electric. OPGW is a technology fusing conventional power cable with optical fibre, which is employed to carry light pulses and electronic signals simultaneously. This is used by electric power companies for transmitting signals for controlling power plants, since optical communication has the merit of being unaffected by electric signals.

Normally, light pulse signals have to be converted back to electrical signals before they can be amplified. Research is underway to amplify light signals directly without converting them to electricity. Progress has been made using erbium-doped optical amplifiers which boost light as it travels along the fibre.[53]

*(e) OEICs (Optoelectronic Integrated Circuits)*

OEICs are devices integrating optoelectronic elements (lasers, LED and photodiodes) and electronic elements (transistors and diodes) on a single substrate to perform complex functions. While optoelectronic technologies are suitable for transmitting information, electronic technologies are suitable for switching and amplifying. Although the earliest form of OEICs appeared around 1980, commercialization has been a slow process, since the integration has been more difficult than expected. Furthermore, it is not clear if the integration of two components leads to a simple chip which is cheaper than the two separately. OEIC was an area assigned to several firms while they participated in the Japanese Optoelectronics national project over the period 1979 – 86. Some firms have temporarily suspended their research on OEICs and have decided to 'wait and see' if there is a suitable application.

The integration of optoelectronics with electronics leads to several advantages. These include high-speed operation, compactness and high reliability. High-speed operation is particularly important in giga-bit transmission systems. The technologies needed for OEICs are those required for optoelectronic key components, and for III – V integrated circuits. Applications of OEICs involve communications and information-processing. Some firms have been working on OEIC in an attempt to overcome the problems posed in increasing the speed of LSI (Large Scale Integration) namely the limited capacity of wiring and large amount of heat generation. By using light as a form of wiring, LSIs could be made faster. However, it would be necessary to lower the power consumption of semiconductor lasers at least to the level of current LSIs in use.

There is a distinct field called 'integrated optics'. This refers to the use of optical waveguides in a suitable material, such as lithium niobate, to carry out optical circuit functions. This is suitable for some functions such as phase modulators, routing switches and splitters.

*(f) CCDs (Charge Coupled Devices)*
CCDs (Charge Coupled Devices) are light-sensing elements used in video cameras, photocopiers and fax machines. Area-type CCDs are used as light-sensing devices in videotape recorders while line-type CCDs are used in facsimiles and copiers. The component generic technologies used to develop CCDs are related to those required to make CMOS ICs. For example, they include crystal growth, etching, photomask, colour filter and microscopic connection technologies.

**Component Generic Technologies**

The key components which have been described above can be developed by application of several generic technologies, such as epitaxial growth. Changing the thickness, structure and the materials of heterostructure formation allows one to obtain semiconductor lasers with different performance characteristics, for example in terms of the wavelength emitted, output power, electrical and optical characteristics and noise. For example, a wide range of wavelengths from the infrared to the visible range may be selected by changing the contents of the aluminium ratios in the active layer. Another point to note here is that different applications have focused on improving different performance parameters. These issues have been discussed earlier. Similarly, adjusting the materials of heterostructure formation allows photodiodes with different performance characteristics to be obtained.

One of the key points in developing LEDs is to improve the efficiency between the LED and the fibre in order to obtain high coupled power. For this purpose, ion beam etching and chemical etching are relevant component generic technologies.

Another set of component generic technologies which are important during the development of components relate to those which are used for measuring crystal defects and examining the degradation behaviour of the components. Such techniques as SEM (Scanning Electron Microscopy), TEM (Transmission Electron Microscopy) and cathodoluminescence are well established. Certain features are similar in both techniques. While TEM provides information about the internal structure of the thin elements, SEM is used for studying the surface or near surface of elements.[54] Simple epitaxial techniques, etching, lithography, and these spectral microscopy techniques have been used to develop semiconductors.

Multilayer formations found in key components such as semiconductor lasers, light-emitting diodes and photodiodes can be fabricated by epitaxial growth techniques. Epitaxial layers are grown on a substrate of GaAs for short-wavelength lasers, or InP (Indium Phosphide) for long-wavelength lasers. For example, the abbreviation GaAlAs/GaAs indicates that a gallium aluminium arsenide layer has been grown on a gallium arsenide substrate. The oldest, and one of the most common techniques, is called liquid phase epitaxy (LPE), in which the compounds to be grown are dissolved in molten gallium or indium up to saturation and where growth takes place by slowly cooling while keeping the substrate in contact with one of the solutions. After growing a layer the substrate is shifted to the next solution for growth in the subsequent layer.

MBE (Molecular Beam Epitaxy) is an epitaxial growth technique, developed more recently than LPE, in which epitaxial films are grown from beams of atoms or molecules by their reaction with a crystalline surface under ultrahigh vacuum conditions.[55] This process allows much thinner layers to be grown compared to those formed using conventional LPE techniques. In MBE, high performance, highly efficient quantum-well-type lasers can be formed. The heterostructure layers could be made as fine as 100A, in other words several atomic layers.

In addition to LPE and MBE, there is also MOCVD (Metal Organic Chemical Vapour Deposition), which is an epitaxial growth technique in vapour form. MOCVD and MBE both have advantages in terms over control of composition and geometry.[56] In 1979, Sony was the first company in the world to produce visible wavelength lasers by MOCVD. Compared to MBE which is complicated to handle and involves the use of toxic gases, MOCVD has been the preferred method for mass production.

In addition to epitaxial growth, other techniques are needed to fabricate key components. Often, a complex structure has to be grown which cannot be formed through epitaxial growth alone. Corrugated patterns can be formed using photolithography and chemical etching techniques.

## 3.4 TECHNOLOGICAL INTERDEPENDENCIES IN OPTOELECTRONICS

Optoelectronic components and sub-assemblies can be combined to form a variety of systems. In optoelectronics, a key component such as a semiconductor laser is little use on its own and needs complementary technologies to fulfill a useful function. For example, in order to function effectively as part of a communication system, a semiconductor laser which would act as a light source needs an optical fibre as a transmission medium, and a photodiode as a detector. In other words, the key components have to work together to form a working system. We outlined how optical communications became a reality through the development of both semiconductor lasers and optical fibres. In Sumitomo Electric, research on gallium arsenide was taking place, before optical communications emerged. We will recall the quotation by Itami and Roehl (1987, p.105): 'The logic of technological interdependence means that the various elements operate on the same level. Unless all are brought up to the level of the strongest one, having a standout technology is of no use.' Firms have to strive to maintain an effective linkage between systems, key components and component generic technologies. They have to choose whether to acquire the components through markets, or by in-house production. In optoelectronics, although the final systems and products at the higher levels of the hierarchy may be different, as one moves down the hierarchy, we find that they are based on certain types of key components, and that these in turn are all based on a range of common underlying generic component technologies.

Another point to note here is that the development of a key component such as a semiconductor laser can be steered in many directions. It can be used in a variety of systems and products. For example continued research and development on a semiconductor laser for a CD pickup enables a firm to move into more sophisticated markets, such as high power lasers for optical storage devices, and other applications such as welding. In other words, technological trajectories can be steered by the firms' strategies.

We will summarize in Table 3.7 an example of how performance parameters of semiconductor lasers can be modified to suit different applications. The left hand column lists parameters, and the direction of change. For example, in the case of the first row, lowering the threshold current has been a very important criterion for semiconductor lasers for CD players, but not so important for high

*Table* 3.7 Changing technological agenda of semiconductor lasers

|  | CD ----▶ | High Power ----▶ | Visible |
|---|---|---|---|
| Threshold current (Lower) | 3 | 2 | 3 |
| Characteristic temp. (Up) | 1 | 2 | 3 |
| Wavelength (Shorten) |  |  | 3 |
| Output power (Raise) |  | 3 |  |
| Long life (Increase) |  | 3 | 3 |
| Single-mode |  | 2 |  |
| Noise (Reduce) | 3 | 2 |  |

3 very important; 2 important; 1 slightly important.
*Source*: Author's interview with company scientists.

power lasers. On the other hand, raising output power has been crucial for high power semiconductor lasers, but not relevant for CD applications.

In this book, LCDs have been separated from other areas and treated as a special case, since this is a case of technology fusion. The underlying generic technologies which are used to develop LCDs include, materials such as TN (Twisted Nematic) and STN (Super Twisted Nematic), and ferroelectric; IC related technologies such as CMOS driver. TFT arrays fall into two types, amorphous silicon and polysilicon. Other technologies include COG (Chip on Glass), surface mounting, photomask, etching, tape automated bonding (TAB), microscopic connection technologies and reliability testing. Since the mid-1980s, a technology to detect faults in the LCD and have them automatically repaired is growing in importance.

## 3.5  CONCLUSION

The purpose of this chapter has been to present an overview of the history and techno-economic features of optoelectronics. Optoelectronics is a comparatively new discipline, and one which is able to perform new functions beyond the scope of conventional electronic technologies. Optoelectronics is a strategically important competence which must be mastered by firms in a variety of sectors in order to remain competitive. It is a prime example of a core emerging technology which finds applications in many sectors of industry, ranging from communications, computers and industrial and consumer electronics to materials and defence. It draws upon aspects of several generic technologies which form the foundation of a range of key components, leading to a wide range of applications.

Optoelectronic components and sub-assemblies combine to form a variety of systems. Optoelectronics offers us the possibility to examine firms' strategies towards maintaining effective linkages between systems, key components and component generic technologies. A key component such as a semiconductor laser can be used in a variety of systems such as optical disks and optical communications. Therefore, a study of optoelectronics-related competence building would enable an understanding to be gained not only of the interlinkages between the three levels of technologies, but also of the firms' strategies in exploiting economies of scope.

In optoelectronics, one can see a clear pattern of innovation emerging in the field, from laboratory discovery in the 1960s, to commercialization and rapid diffusion in the 1970 and 1980s. Finally, we have shown how optoelectronics is an area in which Japanese firms seem to have caught up and in some cases overtaken their Western competitors. This would not have been the case had another choice been made to pursue an alternative competence such as software or AI.

In Chapter 4, the factors affecting the rate and direction of competence building will be examined, based on the model put forward in Chapter 2. At this stage, we shall be focusing on the path dependent-related elements.

# 4 Building Optoelectronic Competence in Firms

The purpose of this chapter is to examine the factors affecting the rate and direction of the competence-building process in firms. In Chapter 2, we put forward a model of competence building. When a radically new, pervasive technology emerges, some firms are able to seize that opportunity to their advantage. Competence building centres in key areas to enhance the core capabilities of firms. Competence building is affected by factors related to path dependence which have direct links with the firm's history, such as previous core businesses, long-term top management strategy, the evolution of the R&D organization, and government policies. It is also affected by factors related to the present, such as management of the interlinkages between systems, key components and generic technologies, and economies of scope. In this chapter, we shall concentrate on the path-dependent factors and organizational routines.

This chapter begins with brief histories of the firms' development, presented in tabular form. For each group of firms, we analyze the main path-dependent factors affecting the rate and direction of the competence development. The second part of the chapter examines the competence-building pattern, during the early, middle, and later phases. Finally, we outline the various organizational routines for sustaining competences.

## 4.1 BRIEF HISTORIES OF THE SELECTED FIRMS

Our research concentrated on seven Japanese and four European firms in the industrial, consumer electronics, communications, electronic devices and materials sectors. As shown in Chapter 3, optoelectronics has affected many sectors, out of which we have chosen the three largest markets on which optoelectronics has had a significant impact. The selected firms have been the leaders in their respective fields, each noted for having a distinguished record of growth and innovation and each having a distinct corporate culture and a style of management.

Furthermore, most of the firms had one element in common: that they perceived optoelectronics as a revolutionary emerging core technology, providing them with a potential competitive edge in transforming their products and production processes.

The main business areas in which the firms have been engaged are summarized[1] in Table 4.1. The key statistics on the firms are summarized in Table 4.2.

The case studies of the ten firms have been summarized in Appendix E.1 to E.10 in tabular form. The Appendix table has a common format for each firm, and the three columns contain the following: the first column lists critical events, including historical events related to both general and optoelectronics matters, and major changes in the strategic direction of the firm; the second column outlines organizational features concerning the evolution of the R&D organization and the structure of the firm; the third column summarizes the role of other agents, including main customers, other firms, and the government. The purpose of the table is to show in a concise manner the evolution of the firms with respect to the three sets of factors. Such a scheme highlights the interaction between the three dimensions and enables us to gain an indepth understanding of each firm's historical development. For example, we can examine the interaction between major changes in corporate strategy and R&D organization. The Appendix table also enables us to draw comparison across firms more easily. The data in the Appendix tables were obtained from a variety of sources including interviews with the firms' scientists and managers, publicly available data such as books, journals, annual reports and brochures, general press reports and the INSPEC database.

Examination of the main business areas of the firms and their histories supports our grouping of the firms into three categories of (i) communications-driven, (ii) consumer and industrial electronics-driven and (iii) other firms. We shall see later in Chapter 5 that competence score analysis using statistical techniques leads to the same grouping. In the next section, we analyze the firms in each group to discuss their similarities and differences with respect to each factor.

Table 4.1 Sales breakdown by business activities in 1985 (units in %)

| | Consumer Elec | | | | ElecSystems/ Devices | ICT | | | Medical Engineering | Cables and Wires | Defence | Other |
|---|---|---|---|---|---|---|---|---|---|---|---|---|
| | AV | TV | Home | Lighting | | Computers/ IT, Comm | | Power/ Heavy Ind | | | | |
| Sony | 63.2 | 26.1 | | | * | | | | | | | 10.7[1] |
| Sharp | | 45 | 22 | | 33[2] | | | | | | | |
| Philips | 26.3 | | 10.3 | 12.4 | 18.1 | 27.8 | | | | | | 5.1 |
| Toshiba | | 33 | | * | * | 35[3] | | 32 | * | | | |
| Hitachi | | 23.7 | | | * | 36.0 | | 40.3 | * | | | |
| Siemens | | | | 3.9 | 4.6 | 14.6 | 16.4 | 42.1 | 8.7 | | | 9.7[4] |
| Fujitsu | | | | | 13.4 | 71.9 | 15.7 | | * | | | |
| NEC | | 7.2 | | | 19.0 | 39.1 | 34.7 | | | | | |
| STC | | | | | 14.8[5] | 54.6 | 25.7 | | | | 4.8 | |
| GEC | | 5 | | | 31.6 | | 12.7 | 32.8 | 8.0 | | *[6] | 9.9[7] |
| Sumitomo Electric | | | | | * | | | | | 73 | | 27[8] |

Source: Annual Reports.

AV – Audio Visual Equipment; Home – Home Electronics; ElecSystems/Devices – Electronic Systems and/or Devices; IT – Information Technology; Comm – Communications; Power/Heavy Ind – Power Systems, Heavy Industrial Equipment.
*Indicates fields in which the firm is active but for which data are not provided.

1 Includes music.
2 Includes industrial electronics.
3 Includes devices.
4 Includes electrical installation.
5 Includes distribution.
6 Rough 30–40%.
7 Includes distribution.
8 Includes brakes, equipment, construction work, hybrid products.

*Table 4.2* Key statistics on the 11 firms in 1989

| Firm | Sales(M$) | Prof(M$) | R&D(M$) | SalesGr |
| --- | --- | --- | --- | --- |
| Sumitomo Electric | 6 293 | 168 (2.7%) | 160 (3.7%) | 6.9% |
| Sony | 15 952 | 525 (3.3) | 1 030 (11.3) | 11.8 |
| Hitachi | 46 387 | 1 344 (2.9) | 2 143 (9.1) | 7.9 |
| Fujitsu | 17 300 | 507 (2.9) | 1 720 (11.8) | 16.4 |
| NEC | 22 339 | 467 (2.1) | 1 884 (10.2) | 15.2 |
| Sharp | 9 122 | 211 (2.3) | 486 (6.8) | 11.3 |
| Toshiba | 27 542 | 865 (3.1) | 1 508 (7.1) | 9.0 |
| STC | 4 267 | 188 (4.3) | 444 (10.4) | 19.6 |
| GEC | 9 620 | 779 (8.1) | 1 000 (10.4) | 8.2 |
| Philips | 26 980 | 597 (2.2) | 2 149 (8.0) | 5.1 |
| Siemens | 32 515 | 845 (2.6) | 3 577 (11.0) | 7.8 |

*Notes*: 1. The units are million US$.
2. The percentages in brackets in the profits column are the % over sales.
3. R&D expenditure for parent companies are listed for Japanese companies.
4. Consolidated R&D expenditure is listed for western Companies.
Prof=Profits. Sales Gr=Average year-on-year compound growth over a 10 year period (1979–89). OECD exchange rate of Y138 was used.
*Source*: *Annual Reports*.

## 4.2 CORE BUSINESSES AND LINKS WITH CUSTOMERS

**Group 1: Communications-driven Firms (Siemens, Fujitsu, NEC, STC, GEC, Sumitomo Electric)**

The firms in this group are similar in that they all have had close ties with the national PTTs.[2] The close relationship with their respective PTTs allowed them to build competences in telephone exchanges, public communications systems and equipment. Having become the main suppliers to the PTTs long before deregulation took place, the companies were able to benefit from the PTTs' aggressive plans in the post-war period to set up a nation-wide communication infrastructure. The special relationship with the PTTs gave an advantage to the firms in the form of technological collaboration and guidance and a steady source of income. In the case of the British firms, as Cawson *et al.* (1987) point out, these private suppliers (Principally GEC, Plessey and STC) were able to marshal formidable economic and political resources to defend and advance their positions vis-a-vis the Post Office under the 'telecom club' regime, even when the Bulk Supply Agreements[3] terminated in 1969.

The PTTs' strategies influenced the way firms planned their activities. For example, the decision of NTT (Nippon Telephone Telegraph) to give up milliwave communications in 1970 had an immediate effect on supplier firms like Sumitomo Electric, which terminated its research programme and transferred researchers to work on optical fibres and semiconductor devices.[4] As mentioned earlier in Chapter 3, most firms also switched the material of the semiconductor lasers they were working on from gallium to indium-based material around 1977–8. Part of the reason why the Japanese firms succeeded in commercializing semiconductor lasers while US firms failed is that the latter did not switch to indium based material, as they considered it too risky to change the material as well as the structure of the lasers.[5] Thus, NTT played a significant role in acting as a technological driver for the Japanese firms. Siemens likewise had a close relationship with the German PTT and with the energy sector.

NTT decided to go ahead with optical communication in 1974. By then, the Japanese companies had already begun to work on optical communications, believing that optical fibres would take over from copper cables. Although Fujitsu was a comparatively late starter, it began work on semiconductor lasers around 1973. After NTT's decision in 1974, Fujitsu decided to concentrate on making optical communications a reality and to catch up with NEC. Before the firms entered the field of optical communication, most of them had accumulated competences in communication, such as microwave communication. Although the latter is somewhat different from optical communication, the underlying technological principles are largely similar, in that both deal with waves which vary in wavelength.[6] Firms were therefore able to build on their existing technological know-how in such techniques as pulse-code modulation, transmission protocol and IC techniques. Cable manufacturers such as Sumitomo Electric had accumulated expertise in techniques such as cable-coating and cable-connection which proved useful for making optical fibres.

Most firms in this group also benefited from having close ties with other public sectors such as broadcasting, electric utilities, energy, transportation or defence.[7] Historical analysis shows that they all benefited from lucrative government contracts until quite recently – for example, GEC with its dependence on the public sector institutions such as MoD (Ministry of Defence), British Gas and British Rail. As the main supplier to BT and MoD, GEC accumulated competences in both civil and military communication such as public, radio and microwave communication and radar and military reconnaissance systems. GEC's areas of strengths are therefore related to its core competences in defence systems, communication and signal-processing.[8]

The close ties of certain firms with electric power companies enabled them to try out optical communication systems on their customers' premises, thus providing opportunities to 'learn by doing'. In the case of Sumitomo Electric and NEC, this took place in the mid-1970s before they began supplying optical communication systems for NTT, so the role of these large corporate customers may have been as important as the PTTs.

The areas of weakness in these firms lie mostly in consumer electronics-related areas such as optical disks and liquid crystal displays (LCDs). As the historical analysis shows, these firms chose not to be actively involved with consumer electronics, although at a certain stage some firms did consider entering the market. Since LCDs found their first applications in watches and calculators, they were outside the scope of these firms. The early application of optical disks was mainly for compact disks which were also thought to be outside the scope of these firms by the R&D managers, so they decided not to enter the market when the initial applications emerged.

Some firms did enter these fields later, and some managed to catch up quickly – for example, NEC with LCDs. For others, however, it has not proved so easy, since competence building requires long-term, continuous effort. Fujitsu's weakness in LCDs may be related to the fact that they have been strong in mainframe computers rather than PCs which require the use of LCDs. In the case of GEC's involvement with LCDs, unlike the Japanese firms which have been developing them for the civilian market, GEC has been specializing in military applications, using LCDs as an instrumentation device for avionic systems. As we saw in Chapter 3, the initial discovery of LCDs was in research funded by the Ministry of

Defence in the UK. Thus GEC's LCD competence has centred on its core competence in defence electronics.

### Group 2: Consumer and Industrial Electronics Firms (Toshiba, Sony, Sharp, Philips)

The three Japanese companies[9] in group 2 have one important feature in common – they were not suppliers to NTT until after its privatization. Unlike the firms in group 1 which have benefited from a long close relationship with the PTTs enabling them to nurture competences in all aspects of communications, these companies in group 2 have not had such an experience, with the result that their capabilities in communications are rather limited in scope. Therefore for these firms, the competence scores[10] (Table 5.14, p.136) show that they are relatively weak in optical communication-related areas. Privatization offered new opportunities for these companies to become suppliers to NTT. For example, all three companies began supplying terminals such as telephones, but in terms of building capabilities in communication systems, they were less advanced. An exception perhaps is Toshiba, which was able to build some competence in optical communications by capitalizing on its close links with electric power utility companies. Like NEC, Toshiba developed trial optical communication networks for electric power companies in the 1970s. Despite the drawback of not being part of the NTT 'family', Toshiba was still able to enter optical communications, by finding applications in areas in which it had prior related technological and marketing competences.[11] Toshiba has since been working hard to catch up, expecting a huge market when 'fibre to the home' takes off.

For these firms in group 2, there has been another area acting as a technological driver for building their optoelectronics competences. For a long time, the main customers for Sony and Sharp included broadcasting companies and individual customers. It was only in the 1970s and 1980s respectively that the two companies began to increase the proportion of non-consumer electronics sales. Thus these two firms have been able to build their optoelectronic competences in areas related to audio visual products such as compact disks, broadcasting equipment and LCDs.

Toshiba's product market has been broad compared to the other two firms, stretching from consumer and industrial electronics and IT to energy, transportation, medical electronics and devices.

Information and communication systems together with electronic devices account for over half of Toshiba's revenues, with heavy electrical and consumer products sharing the remainder.[12] Thus Toshiba's areas of strength reflect its broad business activities. Its optoelectronic competence has centred on areas related to office automation and consumer/industrial electronics.

**Group 3: Evenly Balanced Firms (Hitachi)**

Although Hitachi was part of the NTT 'family' prior to the latter's privatization, traditionally it was a supplier of telephone exchanges and unlike Futjisu and NEC, did not get involved with NTT in developing public communication systems. Lack of experience in public communications systems was the main reason why Hitachi was not able to gain sufficient momentum to build competence in optical communications. After the NTT privatization, opportunities for winning contracts to supply communication systems opened up for companies such as Hitachi, so it has since been trying to build competences in the area. Hitachi's product markets range from heavy industrial sectors, IT, medical electronics, computers and communications to electronic devices. Therefore, Hitachi is the only company whose optoelectronic competence building has been evenly balanced between communications, and consumer/industrial electronics.[13]

The analysis in this section has shown that the firms in each group have had similar core businesses and types of customers. Those with close ties with the national PTTs and the public sectors form one cluster of firms, while the second consisted of firms whose main activities centred on consumer and industrial electronics. Hitachi falls between the two groups, its core businesses and customer links spanning the two domains.

## 4.3   STRATEGY AND THE EVOLUTION OF THE R&D ORGANIZATION

**Group 1: Communications-driven Firms (NEC, Fujitsu, Sumitomo Electric, Siemens)**

As we saw earlier, NEC grew by winning large contracts from the public sector both in Japan and overseas. In 1964, Kobayashi, the president of NEC was influenced by a book written by Machlup (1962) of Princeton University, and came to the conclusion that the future of NEC lay with the 'knowledge industry'.[14] This led Kobayashi to embark on a company-wide effort to shift NEC from being a firm serving the public sector to one led by consumer demand.[15] NEC's top management realized at a very early stage that their heavy dependence on lucrative government sectors could not continue indefinitely, and began taking action more than twenty years ago to prepare for this change. This was by no means an easy task and took much time and effort. Several organizational changes took place gradually over a decade in pursuit of this long-term goal.

The top management of both NEC and Fujitsu designated computers and communications as core business domains in the early 1960s. Subsequent events can be seen in terms of this strategy – for example, Fujitsu spun off its robot division as a subsidiary in 1972 and NEC withdrew from nuclear energy in 1965, enabling them to allocate resources to the two priority areas. In 1977 Kobayashi, who became chairman of NEC in the previous year, announced the 'C&C concept', – a vision based on the integration of computers and communications through merging core technologies.[16] NEC realized that these two domains, which until then appeared unrelated, were interlinked through semiconductor devices and component core technologies, and set about investing in building up technological capabilities in these areas. Although Fujitsu has not apparently defined a corporate identity such as C&C, if one examines the actions taken since the mid-1960s, it is clear that they have been following a similar strategy. In particular, there has been increased emphasis on computers, with the result that their computer business has grown much more rapidly than communications.[17] Their lack of expertise in microwave technology however, may have caused them to lag behind NEC in communications. Another point to note is that as early as 1959, the Fujitsu president, Okada was stressing the importance of core technologies.[18] Given that the notion of core technologies has only become widely discussed in the last few years, Fujitsu's top management seem to have discovered this concept comparatively early, enabling the firm to pursue a policy of investing in core technologies over a long period.

Fujitsu and NEC adopted an SBU (Strategic Business Unit) structure in 1961 and 1966 respectively. SBUs were treated as autonomous units, competing for resources such as capital or R&D expenditure. The shift in organizational structure to one more suited to meeting private demand gained momentum following a top management decision in the late 1960s to transform NEC. It was seen, for example, in the setting up of a comprehensive dealer network for electronic devices which later began to handle more of NEC's various products. A system of matrix management was also adopted. Thus, NEC was able to transform itself successfully to become less dependent on sales to the public sector, which fell from 33.2% of total sales in 1975 to 22.2% in 1980.[19] The organizational structure of Fujitsu consists of three SBUs responsible for computers, communications and devices. Management is highly decentralized with each unit having its own marketing, business planning and R&D planning sections. The level of central coordination by the corporate headquarters is weak compared to NEC.

Sumitomo Electric's president Kitagawa, who was in control of the firm from the 1950s to the 1960s, stressed the importance of developing technological capabilities in-house, rather than licensing foreign technology. This was similar to NEC's strategy. In the 1960s, the top management at Sumitomo Electric pursued an aggressive diversification strategy in order to increase sales of products other than cables and wires, since it was realized that, like any other commodities, profits for these products would diminish over time. A goal was set to increase the non-cable and wire-related sales to 50% over the next twenty years.[20] By 1990, sales of cables had fallen to 53% of turnover and Sumitomo Electric had more or less achieved its goal. It was against this background that Sumitomo Electric made strenuous efforts to assimilate optoelectronics technology, not only to sustain its diversification strategy, but mostly in order to continue supplying NTT, their main customer. Optoelectronics was a competence that they had to develop in order to survive. Top management therefore actively supported the development of optical fibres. During the 1980s top management advocated a corporate identity centred around the notion of 'Optopia', signalling to the world and to its employees that Sumitomo Electric was seeking to become a leader in the field of optoelectronics.

From its early days, Siemens has been a company with a strong technological foundation. The summary of Siemens given in Appendix E.4 shows that the company's growth was based on its deep and broad technological capabilities and its close contacts with public sector customers such as the German PTT and power utility companies. The company's strategy has been to manufacture electrical and electronic goods and systems in both advanced technology sectors and high-margin sectors in mature industries, thereby generating above average growth rates.[21] At one point, Siemens considered entering consumer electronics, but they decided against it, since the margins were not considered sufficiently high.[22] In 1984, domestic sales accounted for 49%, while sales to the rest of Europe accounted for 22%. In contrast to NEC where top management sensed in the 1960s that the future environment would change, in the case of Siemens the realization came a little late – in the mid-1980s – that they would no longer be able to continue to rely on lucrative government contracts. This may have been partly due to the relatively late deregulation of the German PTT, planned for 1995, and partly to the fact that competition in Germany was not as intense as in Japan. Consequently, Siemens did not feel the intensity of international competition until the 1980s. A series of changes took place thereafter, with the company designating the four areas of office automation, telecommunications, semiconductors, and computer integrated manufacturing, as vital to the firm's future. Until the late 1980s, Siemens' organization was more centralized with a large number of management levels. After a major reorganization in 1988, management functions were streamlined. The new organization was designed to increase flexibility and improve the speed in responding to market change.[23] The director of corporate R&D set about forming an R&D strategy based on core technologies in 1988.

Siemens and Fujitsu (which can be considered as Siemens' 'grandchild' since it was an offspring of Fuji Electric which was in turn a joint venture between Siemens and Furukawa Electric), have similarities in the way their organization of R&D has evolved. Sumitomo Electric and NEC also share certain common features (see Figure 4.1).

In all these firms, R&D was originally conducted within separate divisions, since the key technologies used by different businesses were largely independent of one another. At Sumitomo Electric, for example, the R&D unit in Itami conducted work on energy, automobile-related technologies and III–V semiconductors, while the R&D unit in Osaka worked on metals (such as copper wire), and electricity.[24] As the markets evolved and the boundary between different product-markets became more blurred, at some stage a major reorganization of R&D took place in order to conduct research cross cutting the company. This change occurred in 1971 at Sumitomo Electric, in 1965 at NEC, in 1968 at Fujitsu (just after top management designated computers and communications as the

*Figure* 4.1 The evolution of R&D organization at NEC

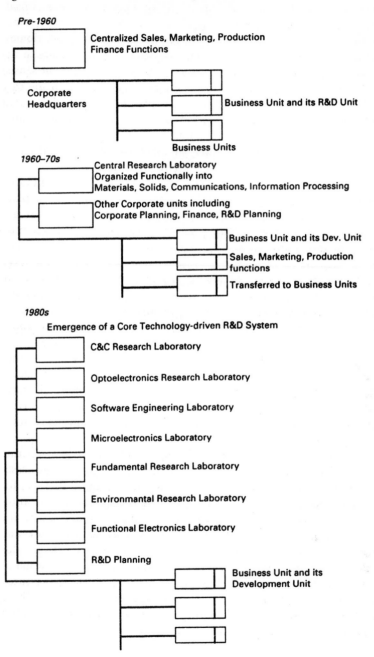

two core businesses), and in 1969 at Siemens. In recent years, both NEC and Siemens have been able to change their R&D organization to one focused on core technologies.

NEC underwent a major exercise lasting two years from 1983 to identify their core technologies. This enabled company scientists to become more aware of how their research affected a range of products. At the same time, it enabled middle managers to plan their research in line with corporate strategies. The exercise proved a useful tool for planning R&D. The central research group is now divided into several laboratories based around core technologies, one of which is optoelectronics.[25] NEC was apparently the first company to form an optoelectronics research laboratory (in 1980), integrating the research done on systems, key components and component generic technologies, thus achieving vertical device integration. Since then, several firms have followed NEC's model, including Hitachi in 1988, and Sumitomo Electric a year earlier, the latter bringing together the research on systems in Yokohama and that on III–V semiconductors at Osaka.

From the above examples, we can see that the top managements of NEC, Fujitsu and Sumitomo Electric were able to detect signs of the changing environment well in advance, and were able to take actions, setting appropriate long-term objectives. These objectives were quickly and effectively disseminated throughout the company.

### Group 1: Communications-driven British Firms (GEC, STC)

A common feature of the two British firms, STC and GEC, emerges if one examines the critical events that they faced. In both cases, major changes occurred on the basis of top management decisions – of a drastic nature – which had a profound effect on the company. Compared to the Japanese firms, which evolved more or less steadily, following clear cut long-term strategies, these two firms have been subject to top management decisions based on much more short-term perspectives. For example, STC's decision to purchase ICL in 1984 was seen as a way of achieving the convergence of communications and computers. However, this was not achieved, and six years later it was decided to sell ICL as STC found that it did not have the resources needed to sustain R&D at the level required.[26] Six years is short compared with the time-span of Japanese company strategies. Finally, the STC decision to accept the takeover bid by Northern Telecom in 1990 had a profound effect in terms of merging the two companies' global strategies.

The organization of GEC has been highly decentralized but under strict financial control imposed by the chairman, Arnold Weinstock.[27] Indeed GEC sees one of their core competences as financial control.[28] The arrival in 1979 of Derek Roberts as Director of Research, caused some change in the organization of R&D. Since then, centrally directed research, which hitherto had been negligible, came to occupy a quarter of the corporate research laboratories' funding.[29] The remainder is funded by the operating companies and external programmes such as ESPRIT. During the 1980s, a research advisory unit was established to assess the research carried out within the three main research centres. After GEC was criticized by the British government in 1985[30] some organizational changes took place to revitalize the company, for example the setting up of a 25 member board modelled on that found in German companies. The operating companies have been performing R&D quite independently, and central coordination has been minimal. There is no corporate planning division at corporate headquarters (see Figure 4.2).

Between 1925 and 1982 STC was allowed to operate with considerable autonomy by its owner, ITT. The company's research centre, Standard Telephone Laboratory, performed outstanding pioneering work in communications such as pulse-code modulation and undersea and optical communications. In general, about a third of the funding is provided by the operating companies, a third by the corporate headquarters and the rest by government contracts. Since Walsh[31] (who used to be at Marconi, the defence electronics subsidiary of GEC) moved to STC in 1985, the organizational routines[32] became similar to GEC's, pushing operating responsibility down to divisions and instituting strict financial control. Such an organizational form is not conducive to competence building, since it discourages investments to build technological capabilities which require a long-term perspective.

In the mid-1980s, both STC and GEC established 'office of the future' divisions, bringing together 1000 people from many different parts of the company. Such drastic measures do not seem to have resulted in much success. The history of GEC is filled with acquisitions and divestitures largely carried out for financial reasons. By concentrating on financial factors, GEC gradually lost its competence in key areas such as semiconductors (for example, it closed down Marconi Elliott Automation in 1971). In optoelectronics, GEC also failed to sustain its competences in key areas such as semiconductor lasers, where it withdrew from volume

*Figure* 4.2   Organizations of STC and GEC

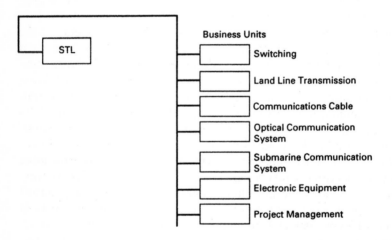

**GEC in 1986**

Hirst Research Laboratory

Marconi Research Laboratory

Engineering
Research Laboratory

3 Corporate Laboratories

Approximately 160
Operating Companies

**STC in 1978**

STL

Business Units

Switching

Land Line Transmission

Communications Cable

Optical Communication
System

Submarine Communication
System

Electronic Equipment

Project Management

production in 1987. Although STC and GEC built up competence in LCDs, the top management of both firms decided not to invest in volume production for applications such as PC terminals, worried at the risks involved, and instead concentrated on niche markets such as defence. If they had made a different choice, they might have been successful in the area of LCDs for civilian applications.[33] The above examples show that the two UK firms tended to have relatively short-term horizons and experienced many changes in strategy. They were also risk averse in relation to new markets.

**Group 2: Consumer and Industrial Electronics-driven Firms (Toshiba, Sony, Sharp, Philips)**

A breakdown of sales over time for Toshiba reveals a declining share of heavy industrial and consumer-related sales against an increase in IT (Information Technology) and communications (Appendix D). Until the 1970s, lucrative consumer products such as TVs and white goods represented a cash cow for the company, supporting the development of the electronics and office automation businesses. The turning point in Toshiba occurred after the oil shock in 1973, when top management had to reassess what was best for the company. With the help of the GE portfolio analysis,[34] they identified IT and electronics as major growth areas. Subsequent actions by Toshiba can be seen in terms of building capabilities in IT, and especially electronics-related component core technologies such as CMOS chips. Optoelectronics emerged at Toshiba against this background, in an effort by the firm to gain competitiveness in IT. The 1980s saw Toshiba announce its slogan of 'E&E' – the merging of energy and electronics. 'Triple I', the combination of Information, Integration and Intelligence, is another slogan that expresses the company's corporate vision. In order to concentrate as many resources as possible on building competences in key components and core technologies, some product markets had to be abandoned, as in the decision by Toshiba to abandon mainframe computers in 1978.

Sharp and Sony are similar in the way that consumer electronics represented their core business for many years. This was the case until ten to fifteen years ago, when their top managements realized that the market for consumer electronics was becoming saturated. As a consequence, they decided to put more emphasis on non-consumer-related areas. As in NEC and Sumitomo Electric, the top managements of these companies were quick to sense the changing business conditions. Both companies realized the importance of building technological capabilities in components and core technologies, and reorganized their R&D to strengthen basic and applied research. Sony stated its goal of becoming an AV&CCC (Audio, Visual, and Computers, Components, Communications) maker and established plans to increase non-consumer products to 50% of the total.

The summary in Appendix E.7 shows that for Sony the turning point occurred in 1983–4 when top management decided to turn it from a captive into a merchant maker of components.[35] Sharp's entry into optoelectronics stems from its vertical integration strategy which began when the company was developing calculators in the 1960s. More recently, Sharp has stated its corporate goal as being 'to fulfil its social obligation as an integrated electronics enterprise built around optoelectronics'. Top management at Sharp regards optoelectronics as one of their Sharp's main core competences.[36]

Toshiba, Sony and Sharp have all been decentralized and divisionalized, with separate business units being the main profit centres. Sometimes an exception has been made, as with divisions such as semiconductors and production engineering, which receive some funding from the corporate headquarters.

These firms have followed a similar path in the way their R&D organizations have evolved. In Sharp and Toshiba, the realization by top management of the need to build competences in components and generic technologies in the 1970s triggered the restructuring of the R&D organizations to form a three-layer model, with a central research laboratory, applications-oriented research laboratories and development units. In both these firms, corporate research laboratories were established in 1961. At Toshiba and Sharp, in 1978 and 1980 respectively, applications-oriented research laboratories began to be established over a ten year period. The corporate research laboratories would conduct research with a five to ten year horizon, while the applications-oriented laboratories would aim at a three to five year time span, and the development departments would be working on current requirements for the coming two to three years (see Figure 4.3).

Kodama (1991) argues that such a three-layer organization allows 'lagged parallel project sequencing' as in the case of DRAM development, an example of a high-technology project. Kodama claims that a three-layer organization allows a highly efficient systematic mode of research and development. Sony have followed a similar path, although they were relatively late in setting up a central research laboratory. One or two small applications oriented laboratories were set up in earlier years, but in 1988, a major change took place with the establishment of an applications-oriented laboratory called the 'Corporate Research laboratory'. This acts as a bridge between the central research laboratory and the development divisions and completes the three-layer R&D structure. The Corporate Research laboratory was located at the headquarters in Tokyo, allowing close interaction between researchers and staff from the business units.

*Figure* 4.3  The evolution of the R&D organization at Toshiba

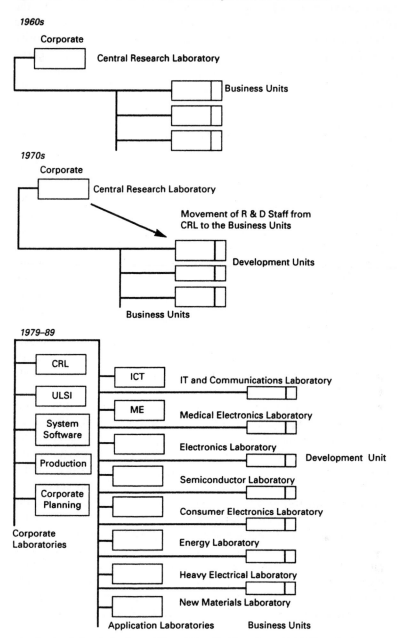

**1960s**

Corporate

Central Research Laboratory

Business Units

**1970s**

Corporate

Central Research Laboratory

Movement of R & D Staff from
CRL to the Business Units

Development Units

Business Units

**1979–89**

CRL

ULSI

System Software

Production

Corporate Planning

Corporate Laboratories

ICT — IT and Communications Laboratory

ME — Medical Electronics Laboratory

Electronics Laboratory

Semiconductor Laboratory

Consumer Electronics Laboratory

Energy Laboratory

Heavy Electrical Laboratory

New Materials Laboratory

Development Unit

Application Laboratories          Business Units

Application Laboratories formed Completing the Three-Layer R & D Organization

By the 1980s, all three firms had reoriented their R&D organization to reflect the emphasis on core technologies. R&D planning units were established to coordinate activities across the firm. The firms are characterized as having a high level of horizontal and vertical linkages. One firm has core technology champions (discussed in section 4.4) and another has a system based around core technology officers and group business officers. In most Japanese firms, Tokkentype systems exist, as discussed in section 4.4.

### Group 3: Evenly Balanced Firms (Hitachi)

Examination of Hitachi's past history suggests that top management strategies or visions have not been as clearly stated as in the other Japanese firms. However, they were implicit, and enabled Hitachi to transform itself from a heavy industrial electrical firm into an electronics firm. The 1960s were seen as a decade in which Hitachi attempted to balance imported technologies with in-house technological developments.[37] The 1970s, by contrast, saw the successful transition into an electronics firm. What is significant about this company is that top management seem to have taken few major decisions which greatly affected the company. Instead, Hitachi evolved steadily with each plant being made a profit centre, instead of the business units as is the case with other firms. This form of organization has been in place since 1960. Authority regarding product development and production planning has been given to the plants. This works well in maintaining a reasonable level of profitability, but may cause a tendency for the company to withdraw early from making products which are not profitable in the short run.[38] It may discourage plants to take risks and make losses for a while as a new business is growing. One point to note is that President Mita in the 1980s recognized the importance of business units and the role of marketing, and tried to increase the power of business units and reduce the power of plants which up until then had been powerful because they were the main profit centres.

Unlike NEC and Toshiba which have taken steps to consolidate their activities, Hitachi has continued to diversify in all areas including consumer electronics, communications, power engineering and industrial electronics. As a result, Hitachi's technological foundation is comparatively broad as well as deep, like Siemens'. The company's optoelectronics activities reflect this. The organization thus evolved around highly decentralized, plant-based profit centres. In 1960, a set of indices were defined to measure the profitability of the plants. Corporate level R&D laboratories performed basic and applied research, which was largely independent of the development activities, which took place in the business units. At a later stage, other applications-oriented R&D laboratories such as the energy and production research laboratories were founded. In the 1980s, the links between R&D and the business units were strengthened, with a transfer of several hundred people from R&D to the business units.[39] At the same time top management put more emphasis on the role of business units, with various changes in the organization. It was only in 1986 that a marketing section was formed within the business units. Profit centres were abolished in the case of certain plants such as semiconductors, which represent key components affecting several business domains. The separate activities in optoelectronics taking place in the firm were brought together in 1988 with the setting up of an optoelectronics research laboratory. A system of Tokken projects was started in the 1970s, allowing horizontal and vertical linkages. The highly decentralized form of organization within Hitachi may make horizontal integration rather difficult. Certainly, core technology champions do not exist. Sometimes, however, ad hoc working groups are formed to monitor the development of certain core technologies such as the group on 'Optical communications in factories' (see Figure 4.4).

## Summary

The examples from the above three cases suggest that Japanese firms have not been subject to the same abrupt changes in top management strategy as UK firms. Japanese corporate strategies have been coherent, with organizational changes and other actions taking place over a long period, and gradually transforming the organization to one better suited for building technological capabilities. In addition, top management in Japanese companies have been quick to sense a changing environment, and has been able to devise appropriate long-term strategies accordingly. The UK firms have relied more heavily on public purchasing and government subsidies than their Japanese counterparts, and have been slower to respond to the changing climate. It is quite common for top management in these Japanese firms to have a technical background[40] which helps them to focus on

*Figure* 4.4  The evolution of R&D organization at Hitachi

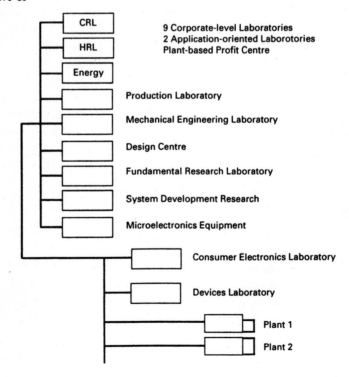

**1960–70**

Three Corporate R & D Laboratories
Plant-based Profit Centres

CRL — Corporate Research Laboratory
HRL — Hitachi Research Laboratory
MEL — Mechanical Research Laboratory

Plant 1 — Business Unit 1
Plant 2 — Business Unit 2
Plant 3 — Business Unit 3

**1970–89**

CRL
HRL
Energy

9 Corporate-level Laboratories
2 Application-oriented Laborotories
Plant-based Profit Centre

Production Laboratory

Mechanical Engineering Laboratory

Design Centre

Fundamental Research Laboratory

System Development Research

Microelectronics Equipment

Consumer Electronics Laboratory

Devices Laboratory

Plant 1

Plant 2

long-term competence building rather than short-term returns on investment. As was pointed out in Chapter 1 a substantial limitation of this study is that it has not examined the role of financial institutions in encouraging firms to pursue a long-term strategy. Understanding and support by top management is essential for building competences. This section has also shown how the optoelectronics-related competence-building process of firms forms part of the overall strategies formulated by top management. In many firms, optoelectronics represented a 'window of opportunity' allowing firms to pursue their long-term strategies.

Top management strategy affects the evolution of the R&D organization, which in turn influences the competence-building process. This section has shown that the evolution of the organization of R&D has been very similar for the firms in group 2, which took many years to establish a three-layer R&D organization. During the 1980s, the organizations have become driven by the goal of developing core technologies with the emphasis on building technological capabilities at the bottom of the three-layer model. While some differences exist among the firms in group 1, in general they show a similar pattern in the way their previously scattered R&D activities were brought together, cross cutting the organization and gradually paving the way for a system focused on core technologies.

Some firms, however, have been highly decentralized – for example, STC and GEC. The organization of Hitachi, the firm categorized as being evenly balanced, was unique, with plants rather than business units as the main profit centres. While the examples of other Japanese firms show how strategy can affect structure, confirming Chandler's (1962) viewpoint, the Hitachi example shows that competence building can be constrained by structure, as Marengo (1991) has argued. However, even in this case, top management recognized the danger and made several organizational changes to overcome the problems associated with a plant-based profit centre approach. Indeed, there was a major reorganization in 1992, abolishing the old system altogether, except in the heavy industries business unit.[41] In several firms, the organization was further changed to integrate devices, component generic technologies, and systems, thus effectively integrating the R&D activities vertically, facilitating interlinkages between systems, key components and component generic technologies. This section has highlighted that forming an organizational structure which is conducive to competence building takes a long time, and requires a coherent top management strategy.

## 4.4 PATTERNS OF COMPETENCE BUILDING

**Initial Phase**

If one examines the motives of firms entering the field of optoelectronics, in most cases one finds that the research was begun spontaneously by researchers who thought it represented an exciting new field of potential importance to the company. Generally, in the Japanese and the European firms examined here, the research began in an ad hoc manner with a small team of people, without first going through a formal strategic planning process. Several researchers commented that it was more like pursuing a dream.[42] In the early 1960s, important discoveries were being made in optoelectronics (e.g. in relation to lasers) and in some firms there was a 'band wagon' effect. In most firms, when they started research, they did not have a clear application in mind. For example, when research on III–V semiconductors was started at Sumitomo Electric, researchers wanted to discover what kind of a device could be made with gallium arsenide, employing the property of the material of being much faster than silicon. The material, which was later used in key components for optical communications, was not at first considered for that application. Eight to nine years later, when new developments in optical fibres began to take place, the application of III–V semiconductors to optical communications emerged. In other words, once the company had invested in building technological capabilities in key generic technologies such as gallium arsenide, it was later able to use this expertise to develop key components in optical communications.

In Toshiba, researchers were interested in the application of lasers to holography in the early 1960s, soon after lasers were discovered.[43] They thought it might be possible to use it as a memory device for information-processing. Thus, the curiosity of researchers can play an important role in the initial phase of competence building. The initial discovery in the laboratory often leads to awareness on the part of top management and structural adaptation. Some firms have been able to create an atmosphere in which researchers are encouraged to tackle new challenges and to take risks without being afraid of failure. In such firms, the organization tends to be loose and informal, often with frequent organizational changes taking place. This tends to be common in Japanese firms.[44] Feedback mechanisms exist which reward the initial search process by researchers with support from top management.

In other cases, firms entered the field of optoelectronics to meet specific internal demands from other divisions. Sharp's strategy of pursuing vertical integration when they were developing calculators is an example: they entered optoelectronics not through a formal strategic planning process, but rather because it was necessary for developing their own products. At Sumitomo Electric, key components began to be developed, following requests from the systems division. NEC entered LCDs late for a similar reason. At NEC there was a large and growing PC business which required LCDs. Furthermore, NEC considered the time was right to enter LCDs when the technology reached a point where it was necessary to use TFT (thin film transistors), an area which would draw upon NEC's existing expertise in IC technology.

The case of Sony may be slightly different in the sense that here the decision to develop CD players was set by top management. This meant it was necessary to develop optical storage technologies. However, in this case, research on MOCVD (Metal Organic Chemical Vapour Deposition), which became a component generic technology, had been taking place before the CD project was launched, so Sony did not start from scratch. Work on semiconductor lasers started in 1975 using LPE (Liquid Phase Epitaxial) growth technique, which suffered from low yield, and was therefore unsuitable for commercialization. In 1978, the team switched to MOCVD since there was already MOCVD equipment in the laboratory which had been used in the 1970s for a research project on MOSFET (MOS Field Effect Transistors) but which had been left idle after the project was terminated. The research on MOCVD which became a leading edge technology for Sony began spontaneously without going through a formal strategic planning process. At the end of 1979, room temperature oscillation was achieved for the first time, and the count of MOCVD equipment was increased to 3. By 1981, although semiconductor laser production using LPE technique was transferred to production, it was becoming more apparent in the CRL (Central Research Laboratory) that MOCVD was superior. In 1984, a decision was made to switch all production from LPE to MOCVD. It was a bold move but it turned out to be the right decision. Being able to select good technologies is a competence in itself. Having switched to MOCVD, the performance of Sony's semiconductor lasers improved to a point where they were able to use internally developed ones in their CD players. They had already accumulated competences in other key technologies, such as mechatronics, lithography, and digital signal processing. In the case of GEC's LCD, the decision was again made by top management.

In the final category of firms, the decision to enter optoelectronics was made in order to enable the firm to maintain links with its main customers. As we saw earlier, when the national PTTs decided to go ahead with optical communications, there was a group of companies already closely linked to the PTTs as suppliers of communications systems. For these companies, investing in optical communications was a vital step to maintain customer links. These firms had little choice. If Sumitomo Electric had not developed competence in optical fibres, it would have lost its most important customer, NTT. If Fujitsu had not invested in optical communications, not only would it have affected its ties with NTT, but its competitiveness vis-a-vis NEC would also have been seriously eroded. Deciding *not* to invest in building competences may be more risky than investing in them.

**The Competence-building Phase**

During the competence-building phase, there are several issues which are worth exploring: the nature of the competence-building process, how firms set about building competences, to what extent the process was based on in-house effort, spontaneity as opposed to explicit strategic planning, and the extent to which government played a role.

Since optoelectronics was a new discipline, and there were no existing optoelectronic specialists, the researchers came from various backgrounds including electrical engineering, physics, computing and chemistry. Some people had worked on microwave and milliwave communications, or semiconductors. In most Japanese firms, the core competences were built in-house without relying on external sources such as universities, government research laboratories, and other firms. This was true also for European firms. However, there were a few exceptions. Sumitomo Electric acquired the business unit responsible for III–V semiconductors of a medium-sized firm called Nikkei Kakou in the mid-1970s, hiring half a dozen specialists on epitaxial growth.[45] The company had been in the process of diversifying into key components, starting from materials and optical fibres. Through the GEC/Siemens joint bid for Plessey in 1988, GEC was able to re-acquire R&D expertise in semiconductor lasers which it had earlier lost.[46] In order to start research on LCDs, GEC hired Dr Hilsum who had done pioneering work on these at RSRE.[47] These cases seem to be exceptions, however, and the major part of competence building has apparently been achieved through in-house efforts. Even in the case of Sony and Philips, the two companies had been working independently for five or six years before they began to collaborate, so in-house efforts were again crucially important. Nevertheless, as we shall discuss in the next section, technical guidance from the national PTTs such as NTT played a significant role in steering technological developments for some of the firms classified here in group one. Collaboration with NTT and other firms allowed firms such as Sumitomo Electric to develop indigenous Japanese technology to produce optical fibres. Universities have played a minimal role in Japan in helping firms to build competences in this area. Most firms suggested that they have been giving universities contracts to do research mainly to maintain links for recruiting the best students to their research group. This is consistent with the work of Nelson and Levin (1986). In the case of LCDs, one firm mentioned that they have been conducting joint research with the supplier of liquid crystal materials. In section 4.5 the role of government policy will be discussed. In Japan, MITI did not play a central role in helping the firms to build optoelectronics-related competences, although it was effective in other ways. In contrast, the MoD and BT played a major role as a technological driver in supporting the British firms.

In recent years, firms which have been competitors in other areas have started to collaborate on their own initiative. An example is the joint venture formed between Toshiba and IBM in 1988 to produce large LCDs. The motives behind this deal are obvious; IBM wants access to Toshiba's LCD production capabilities and Toshiba wants access to IBM's huge market for PCs which could be fitted with LCDs. In other words, this type of joint venture is based on the exchange of complementary assets.[48] In 1991, seven competing cable companies including Sumitomo Electric established a joint R&D centre to conduct research on environmentally-related issues.

Some firms were early starters in optoelectronics but failed to sustain the competence-building process. In the case of Siemens, although the researchers began early in 1963 to work on semiconductor lasers and succeeded in developing them, they soon gave up trying to make them commercially feasible.[49] They shifted their resources to develop LEDs and eventually succeeded in this. Having stopped the work on semiconductor lasers for several years, Siemens restarted several years later but the gap had an adverse affect on its competence (see Table 5.14, p.136, which shows mediocre rating). Similarly, at

NEC, the rather poor rating for optical disks stems from the fact that research on optical disks was terminated for four or five years after the oil shock in the 1970s. The company restarted to work on optical disks when the possibility of using them for computer storage applications arose. However, stopping research for several years may have caused a setback, disrupting the technological accumulation process. Continuous efforts are essential for competence building.

In other cases, the research reached the stage of commercialization, but firms failed to continue with volume production and gave up making key components such as semiconductor lasers, since they could not keep up with the fierce competition (for example, Hitachi's semiconductor lasers for CDs were launched in 1986, but abandoned two years later and GEC gave up producing semiconductor lasers in 1987). In a few cases, competences were lost when the company disposed of a business unit (GEC lost its expertise in optical fibres in 1987, as did STC when it sold its optical fibre division in 1991). As competence building is a long-term process requiring a great deal of trial and error and much perseverance, firms which have abandoned production have generally failed to sustain their competences. These cases highlight the failure to make long-term commitments representing myopic systems as pointed out by Pavitt and Patel (1988) as opposed to dynamic systems. On the other hand, some firms were comparatively late starters who nevertheless managed to catch up with the early starters, as in the case of NEC with LCDs and Fujitsu with optical communications. These firms were able to capitalize on their accumulated generic component technologies such as semiconductor fabrication, lithography and epitaxial growth, and were able to mobilize resources across the firm through top priority projects to accelerate the competence-building process.

During the competence-building phase, some firms have been able to take advantage of opportunities which have come about by chance. Competence building has therefore sometimes been accompanied by a series of spontaneous, haphazard events. For example, we have noted above how research at Sony on MOCVD, which became a core technology for developing semiconductor lasers, started by chance: there was already MOCVD equipment in the research centre, which had been left standing idle following its use for a project on MOSFET in the early 1970s. Consequently, research began without first going through a formal planning process of assessing all the alternative technologies. When the time came for the systems team which was developing CD players to decide which makers' semiconductor lasers to use, the internally produced ones had to be ruled out because they suffered from astigmatism. Their product having been rejected, the semiconductor laser team made strenuous effort to make improvements. The solution to the astigmatism problem was found by chance when an annual open day of the research laboratory was held and the semiconductor lasers were displayed. Someone from the business unit saw the display and suggested a way in which the astigmatism could be corrected by inserting a lens in an oblique position. This changed the research agenda of semiconductor lasers, enabling Sony to catch up with Sharp after a few years. Hence, in this instance spontaneity and learning played an equally vital role as strategic planning.

In section 4.3 the importance of long-term strategies and top management understanding to support the competence-building process was discussed. Strategic planning plays a central role in supporting long-term research and in setting up organizations conducive to competence building. In most firms, once top management understood the potential of optoelectronics, and the need to build competence in the new field, substantial resources were allocated and changes in the organization took place, such as the establishment of an optoelectronics research laboratory.

Table 4.3 summarizes the general trend in the competence-building process. The horizontal axis represents the phases of competence building. The number of asterisks denotes the relative importance.

*Table* 4.3   Summary of competence-building mode

|                  | Early | Mid | Late |
| ---------------- | :---: | :-: | :--: |
| In-house         | ***   | *** | ***  |
| National project |       | **  | *    |
| Universities     | *     | *   | *    |
| With other firms |       |     | *    |
| NTT              | **    | **  | *    |
| MoD, BT          | ***   | **  | *    |

*** Very Important.
 ** Important.
  * Minor importance.

**Sustaining Competences**

Even when competences have been built, it is all too easy for firms to lose them unless they are continuously upgraded. In Section 4.3 on the evolution of R&D organization, we saw how some firms had effectively transformed their organization to suit the competence-building process. In such organizations, there were effective horizontal linkages allowing good communication across different units. There were also effective vertical linkages, allowing good communication across systems, key components and the component generic technology units. Some firms have developed organizational routines[50] for the purpose of achieving vertical and horizontal integration.

**Organizations Based on Vertical and Horizontal Integration**

Most of the Japanese firms studied have a system of 'Top priority' (Tokken)[51] projects allowing them to mobilize resources across the organization to develop particular core technologies and products. The Tokken system allows a form of vertical linkage since it usually consists of several project teams including people working on systems, components and generic technologies. At the same time, the Tokken system provides a mechanism for horizontal linkage, bringing together people from different parts of the organization as well as those working on complementary technologies. Tokken-type projects are monitored by the decision making group, including top management and the divisional and R&D directors, who typically meet about once a month. These projects are given the highest priority in terms of human and financial resources. Such a system is a highly efficient way of developing key technologies or core products. The number of Tokken projects in progress in the sample of firms ranged between five and twenty. Organizations which do not have a Tokken system seem to be more likely to suffer from duplication of effort in different parts of the organization, resulting in wasted resources and longer lead times. Among the firms visited which had Tokken systems, for most of the sub-fields which were rated as being 5 (i.e. areas where the firm was most competent) by the firms, the research was carried out largely through Tokken-type projects.[52] Some firms distinguish between two types of Tokken projects. Firstly, there are the special projects which are given priorities in terms of human and financial resources, to expedite product development. Secondly, there are projects which are aimed at developing core technologies across divisions.

## Core Technology Champions and Other Methods

A frequently used mechanism for improving the effective horizontal linkages in some Japanese and European firms is core technology 'champions'. In this case, core technology champions are responsible on a daily basis for the overall optoelectronics activities, liaising between divisions, manufacturing and the research centres, clarifying their respective responsibilities, and coordinating the research programmes. The champion is in charge of monitoring world-wide trends in optoelectronics, attending conferences to meet academics and researchers from other companies and to discuss progress on equal terms with such people. The champion is responsible for determining both space and time horizons. In other words, the former refers to the task of coordinating activities taking place in different locations concurrently. The latter means that the champion coordinates activities taking place in basic research spanning a ten year period, applied research spanning three to five years and development spanning approximately a year. Often a core technology champion has an unofficial assignment but others in the organization know who the champion is and he/she is widely acknowledged to be playing the role. In some companies, the tenure of a core technology champion is fixed, say for two years.

In some companies, the core technology champion works closely with a partner who is the business manager, the former being responsible for the core technology, the latter for the business unit.

Some Japanese companies also have a system involving a 'Gishichou', or technological chief, whose status is equivalent to that of the SBU manager. Within a SBU, the manager and the Gishichou act like partners with different responsibilities. While the SBU is responsible for the overall activities of the business unit, the Gishichou concentrates on the technological aspects. The Gishichou would coordinate the development of technologies across R&D centres and business units, or look at technical standards, education and training, or perhaps act as a technological gatekeeper as discussed by Allen (1970).

In order to improve horizontal linkages, most Japanese firms hold open days of R&D centres about four times a year at which there are frequent interactions between the R&D laboratories and business units. During regular working days, frequent informal interaction between the R&D and the business people is encouraged. In some firms, study groups for certain core technologies are organized by the technology planning division, enabling interaction between people working in different units.

At Sony for example, research on advanced optoelectronic materials such as semiconductor lasers and optical waveguides is carried out in the central research laboratory, while applied research on optical disks, optical communication and sensors is carried out in the corporate research laboratory. Strong links exist between the development teams in the business units and the corporate research laboratory. It is quite common for an engineer in the business unit to borrow a desk in the corporate research laboratory to acquire new skills. In turn, a close link exists between the corporate research laboratory and the central research laboratory.

Competences are not only held in organizational routines but in other ways. Some firms have been putting effort into developing online databases which hold information on technologies, components and products which have been developed or undergoing development. These databases are accessible from employees in the company, improving the spread of know-how.

### 4.5   THE ROLE OF GOVERNMENT POLICY

**Japan**

Until 1948 when NTT (Nippon Telephone Telegraph) was founded, NTT's research laboratory and MITI's[53] research laboratory ETL (the Electro Technical Laboratory) belonged to the same establishment. The two laboratories became separate entities when NTT was established and since then, there has always been some rivalry between them.[54] During the mid-1970s, NTT and the companies belonging to the 'NTT family' tried to combine their efforts to commercialize optical communications. In Japan, there is generally a clear boundary between the areas of responsibility for individual ministries: for example, the Ministry of Post and Telecommunications is responsible for the communications sector, while MITI oversees the computer and electronics industries. Because communications fell outside its domain, ETL watched the NTT-led developments with some apprehension since it wanted to do something similar.[55]

ETL finally managed to persuade MITI to collaborate, and in 1978 a government-funded national optoelectronics project began to be formalized. It was called the 'Optical Measurement and Control System' project. By the mid-1970s, some companies such as NEC and Toshiba had already reached the stage of running optical communications trials for NTT and for the electric power companies. Optical fibres and other key components used in optical communications were by that stage close to commercialization. In other words, companies had been able to build optoelectronics competences without government intervention. This is in marked contrast to other technologies such as computers and VLSI chips where government intervention and national projects apparently played an important role.[56]

When MITI's coordinator of the project first started to make plans, it was not obvious where they should be focusing their efforts, since there was little research left to do.[57] It was decided that several of the companies which were leading in the field of optoelectronics should be asked to join the project. Usually MITI selects companies to join national projects by checking certain criteria, including the company's technological capability (the higher the better), how many papers they have published (the more they have published the better), and how much equipment they have in the area, since MITI will have to pay less if they are better equipped. The idea is that in MITI-led national projects the best companies are selected to join the project. Using these criteria, MITI selected five companies to be in charge of five sub-systems; the companies were NEC, Fujitsu, Hitachi, Toshiba and Mitsubishi Electric.[58] It was decided that each would concentrate on a different type of key component.[59] NEC would develop a semiconductor laser emitting light of short wavelength; Toshiba would develop a laser which would emit light of several wavelengths simultaneously; Hitachi and Fujitsu would work on different types of OEICs (Optoelectronic Integrated Circuits – the term was invented for the project), these being devices integrating optoelectronics with ICs. After the agenda had been decided, Toshiba began research on short wavelength semiconductor lasers, the topic supposedly assigned to NEC. One effect of national projects would thus appear to be to spur activities in other firms which were not allocated responsibility for those activities by MITI.[60]

Shortly before the project was launched and while the final plans were still being drawn up, Sharp expressed a desire to join. However, since the coordinator did not want to change the plan, Sharp's application was rejected.[61] Since Sharp was not among the potential participants chosen in the first place, it seems that they did not satisfy MITI's selection criteria. When they turned them down, MITI suggested that Sharp work on semiconductor lasers for compact disks. To what extent MITI influenced Sharp in determining its course of action is not obvious from this incident. Sharp had, after all, been a pioneer in the field of optoelectronics dating back to the 1960s, so it had already accumulated expertise in the component generic technologies. What is clear from this incident is that, despite being unable to participate in the national optoelectronics project, Sharp was the first company to commercialize VSIS-type[62] semiconductor lasers for optical disks (in 1979) and it has since maintained a leading position.

Although Fransman (1990) argues that 'It is likely that in the absence of government intervention there would have been significantly less research in the field of optoelectronics', his analysis does not examine the competences accumulated in the firms before the project started. A similar critique can be applied to the work by Sternberg (1992) who argues that the failure of the US government to adopt a coherent industrial policy led to a decline of competitiveness of the US photonics industry; in contrast, Japanese agencies were able to respond with deliberation to the rise of an emerging revolutionary technology, enabling Japan to take the world lead in optoelectronics. The firms that participated in the national project had, by the time the project started, already accumulated competences in optoelectronics. Hence, the national project does not seem to have played a critical role in building optoelectronics competences in these firms. However, there were undoubtedly various secondary benefits. For some firms, the fact that they were chosen by MITI was symbolically important. The project enabled such firms to venture into an area in which they might not have become been involved on their own.

MITI's role can be seen from another perspective. MITI and the Ministry of Post set up a trial of an optical communications system called HIOVIS (Highly Interactive Optical Visual Information System) in 1978, in a town near Nara. The system linked 158 homes via an optical network. The demonstration project acted as a major showcase, enabling people to try out optical communications and see what they could do with it. Setting up such a system catches the attention of people and of the media, promotes the use of new technologies like optoelectronics and spurs activities across companies.

Another initiative involved the industrial association, OITDA (Optoelectronic Industry and Technology Development Association) which was founded by MITI in 1980. Its aim was to draw together all those active in the sector to do the following: (1) collect data annually on production and market growth; (2) coordinate standardization issues; (3) organize seminars and conferences; and (4) commission studies into long-term market trends including activities abroad. The association seems to have been beneficial and played a catalytic role, for example by speeding up standardization issues and encouraging people from competing firms to meet and share information. Consequently, in the absence of MITI's efforts, optoelectronics might have developed more slowly and in a more fragmented manner. On the one hand, the national optoelectronics project began rather late, just as various components and systems were about to be commercialized, and it made little contribution to helping firms build up their competences. On the other, it may have been important in other aspects, signalling to the world that optoelectronics had been chosen as a national project and, spurring activities across firms which were not chosen to join.

## Europe

We saw in Chapter 3 how the Ministry of Defence (MoD) played an important role in initiating LCD research in the UK by funding collaborative research between RSRE (Royal Signals and Radar Establishment) and Hull University. Similarly, firms which were engaged in developing components used in optical communication benefited from support first from the MoD's Directorate for Components and Valve Department (DCVD), and then from the British Post Office (later reorganized as BT). Like NTT in Japan and ATT in the US, BT has been the major driving force in Britain behind the research, development and technological exploitation of optoelectronics. When optical communications began to look feasible, the British Post Office enthusiastically supported it, through its own research as well as by influencing other government agencies notably the MoD to support research.[63] When Corning produced an optical fibre with an attenuation of less than 20 db km⁻¹ in 1970, British Post Office had also initiated a programme to develop optical fibre communication technology with its suppliers.[64] STC benefited from the MoD funding in its development of lasers. In recent years, as support from the MoD and BT has declined rapidly, the SERC (Science and Engineering Research Council) and DTI (Department of Trade and Industry) have increased their contribution to medium-term, long-term and collaborative research through JOERS (the Joint Optoelectronics Research Scheme).[65] In addition, FOS (the Fibreoptics and Optoelectronic Scheme) offered strategically targeted assistance for product development between 1981 and 1987. STC, for example, benefited from FOS support for developing undersea transmission systems. UK and European companies are cooperating in the European Community RACE project and on certain important component technologies in the ESPRIT programme.[66]

The discussion above leads us to the following conclusion: the MoD and the British Post Office played an important role both as a technological driver, and as a funder of the research, especially in the early stages of technological development. They helped the firms build competences (as shown by STC's growing strength in undersea optical communications). However, in other areas such as LCDs, although British firms have been successful in niche markets, they have not been able to convert their early lead into a larger market for other applications such as PCs. This was mainly due to the firms' policies of risk aversion discussed earlier. Furthermore, unlike in Japan, where optoelectronics emerged as a core technology in the presence of demonstrator projects and the creation of an industrial association (OITDA), such stimuli were missing in the UK.

### 4.6 CONCLUSION

In this chapter, we have discussed various path-dependent factors affecting the rate and direction of competence building. A historical analysis of the firms led to a classification into three groups according to their previous core businesses and type of customers. The first group included firms with close ties to national PTTs in supplying communications systems, while the second group consisted of firms focusing on other core businesses such as consumer and industrial electronics. Hitachi is the only firm which is fairly evenly balanced between these two. Top management strategy affects the evolution of the R&D organization. Understanding and support by top management is essential for building competences. Japanese firms' strategies have been coherent, enabling organizational changes and other actions to take place gradually over a long period. Such strategies are better suited to building technological capabilities. Also, top managements of Japanese firms have been quick to detect changes in the business environment, and have been able to devise appropriate long-term strategies.

The second part of the chapter examined the competence-building pattern. An analysis of the initial entry into optoelectronics showed that in many firms, work was begun spontaneously by company researchers without knowing exactly what they could do with it. In other firms, the decision to become involved was due to internal demand, and in some cases it was a necessary step for maintaining links with the main customers. Yet, whatever the initial mode of entry, top management support is essential for competence building. The chapter also outlined the methods used by firms to sustain competences. Various forms of organizational routines centred on horizontal and vertical integration were discussed.

Our analysis of the role of government policy showed that NTT in Japan, and BT and MoD in the UK all played significant roles as technological drivers. MITI's role as a technological driver has been minimal, although it brought about other benefits such as encouraging activities across other firms, and paving the way for the market to take off. Until now, we have mainly focused on path-dependent factors affecting competence building. Chapter 6 is dedicated to the current forces, namely vertical interlinkages and organizational learning, which influence competence building. In Chapter 5 we shall explore the possibility of assessing competences through the use of quantitative methods.

# 5 Empirical Results – the Classification of Competences

The previous chapter examined factors affecting the rate and direction of competence building in firms. We discussed the competence-building pattern based on our analysis of the case studies of 10 firms. Optoelectronics-related competence building is an example of the marriage of a revolutionary technological opportunity with a firms' cumulative competences. Firms view optoelectronics from different angles, and the areas on which they focus are shaped to a large extent by their previous accomplishments and accumulated competences.

This chapter begins by outlining how it might be possible to measure competence by using bibliometric[1] data on scientific publications, US patent data and interview data. We discuss the strengths and weaknesses of the methods employed. The results of the INSPEC bibliometric analysis are presented to show the emergence of distinctive trajectories. After setting out the results of the US patent analysis and interview data, we then undertake various statistical tests to examine the validity of the different measures. Competence is evaluated for each firm using a combination of the different measures. Having identified the areas of strengths, we group the firms according to their common strengths and weaknesses. Other statistical techniques are used in parallel to validate the grouping process.

First we will review some previous empirical work on measuring competences.

## 5.1 PREVIOUS EMPIRICAL STUDIES AND THE METHODOLOGY PROPOSED

While relatively little empirical work on competences has been published to date, contributions emerging from several authors will be discussed here. Kandel *et al.* (1991) examined the technical competence of a firm by measuring the expertise profile of the scientists and technologists working in it. While this is one way of measuring competence, it has certain limitations. The main limitation is that the know-how held by individuals is an input variable to the learning process. In their analysis, accumulated competences are measured by the number of man-years spent on such accumulation process. Their technique rests on the assumption that every individual has the same learning capability, an assumption which cannot be true. As Hedberg (1981) points out, 'Although organizational learning occurs through individuals, it would be a mistake to conclude that organizational learning is nothing but the cumulative result of their members' learning'. It is not the individual ability that counts, but the mechanism by which the individuals can be mobilized to take collective action.[2]

Thus, what is needed is an output measure of the learning process – that is, a way to assess the capacity to integrate different streams of skills and the know-how held by engineers and technologists. While this is an output of the learning process, we must point out that in economic terms it represents an input to the production process.

In their analysis of technological diversification (which they defined as a firm's expansion of technological competence), Granstrand and Oskarsson (1991) attempted to measure competence by counting the qualified engineers of various types employed by the firm. This approach was adopted on the assumption that (a) these people are the prime carriers of technological competence and (b) in general, their area of work is close to that in which they received their professional education. We believe that a similar criticism outlined above for Kandel *et al.* would apply to this method. In addition, point (b) would not be a reasonable assumption in Japanese firms, where extensive, in-house on-the-job training is important, so that a worker's original qualification might not have a very direct link with the job content.

The studies mentioned above on assessing competences have centred on the measurement of knowledge input variables, such as the know-how held by engineers and technologists. While this is one way of assessing competences, as we have argued, competence is required to integrate the various streams of technologies, to grasp technological opportunities, and to mobilize resources effectively across the firm. In this study, we shall attempt to measure technological competence within the firm by analyzing two types of output data of the learning process, namely patent data and publication data. In addition, I conducted intensive interviews with scientists and R&D managers working in firms to provide a qualitative account on how optoelectronics evolved within the firms and to obtain their own assessment of the firms' competences. The research focuses on an important technical field considered by many firms as a core technology, namely optoelectronics.

## 5.2  NATURE AND SOURCES OF DATA

### Complementarities in the Data

Various types of output indicators such as patent, bibliometric and technometric[3] (technology specification measures) data have been used to measure innovation trends as well as the strengths of countries (or companies) in selected technological fields.[4] An in-depth analysis of the industrial robot technology in Sweden, West Germany, Japan and USA was conducted by Grupp, Schmoch *et al.* (1990) using patent, INSPEC, technometric and trade data. The authors argue that robotics is a highly complex field, characterized by a short commercialization process, but which is nonetheless amenable to quantitative analysis. One of the major conclusions drawn was that patent literature is as important as scientific papers in providing access to the latest R&D results. The model of phases in the innovation process adapted from Grupp's (1989) model is shown in Figure 5.1.

While technometrics reveal trends in innovation and imitation, and foreign trade data point to trends in diffusion, the phase of competence building that we are interested in is best covered by scientific

*Figure* 5.1  Phase model and output indicators

| Phases | Indicators |
| --- | --- |
| Knowledge | |
| Research (Basic) | Papers |
| Research (Applied, Strategic) | Papers/Patents |
| Industrial Development | Papers/Patents |
| Innovation and Imitation | Technometrics |
| Diffusion | Foreign Trade with R&D Intensive Products |

*Source*: H. Grupp (1989).

papers and patents. During the competence-building phase, especially in the basic and applied research stages, companies will publish papers when research bears fruit.[5] However, this does not necessarily mean that research will lead to successful commercialization. In some cases, it might bring about successful returns on investment, but in others, it might not lead to direct commercial benefits. However, the know-how gained by the researchers will normally remain and form part of the competence accumulated. In other words, it may be a measure of innovation potential. Therefore, we can say that scientific papers are probably useful indicators to measure the accumulation of 'know-why' technological competence, although they might not be good for measuring innovation processes.

The propensity to patent is different from the propensity to publish papers. A research project for which a paper has been written might not be patented if the area is not patentable (for example, if it is basic research or some types of software) or if the firm decides not to patent.[6] Therefore, the use of scientific papers and patents may in some cases be complementary, but the two may also be supportive of each other. In other words, in some cases there might be a one-to-one correspondence between papers and patents, but in others they may reveal strengths in different areas. Publications should be a more sensitive indicator of technological 'know-how' than patents.

It has been suggested by some analysts that measure of foreign patenting in a given foreign country provides a reliable reflection of international distribution of innovative activity than domestic patenting.[7] The use of foreign patent data equalizes the costs of patenting from whatever country. Furthermore, there is evidence that innovating organizations are more selective in their patenting in foreign countries.[8] The importance of using US patent data has been discussed by Pavitt (1986). It is argued that firms in general apply for patents in the countries where they expect their biggest markets. Foreign patenting in USA is likely to be an accurate reflection of the international distribution of technological activity.

At this stage, we shall explain the patenting process of Japanese firms.[9] If an employee in a company prepares a paper for publication, it is checked by the various levels of management, from the section head and the department head, up to the director of the R&D laboratory. If permission is given for publication, a patent is filed in Japan. When this has been done, a firm can wait for up to one year to see if a patent should be filed in the US, since the US patent office acknowledges the patent priority date filed in Japan. After one year, if the firm decides to file for a patent in the US, it then takes approximately two years for it to be granted. Hence, we would expect to see a time lag between publications and granting of US patents of approximately three years. The final stage in this study is to ask the firms' scientists and R&D managers when they assess the firms' competences, to bridge the gap between the reality and what is revealed by the indicators. In total, 11 firms were chosen in this study.

Although the use of product sales data as an output indicator was considered, it was decided not to pursue this for three reasons; first, since the study focuses on technological competence, it would not be possible to separate the contribution of technological and marketing competences from product sales data; secondly, since optoelectronics is generic in nature, having diverse applications, competence in one sub-field may be related to several products; thirdly, there is the difficulty of obtaining reliable product sales data since firms are often unable to provide product sales data broken down into detailed categories. There would not be a one-to-one correspondence between the product data and the classification into sub-fields used for patent data and scientific publications data. It was judged that the use of patent, publication and interview would be sufficient to enable the dynamics of competence building to be analyzed.

**Data on Papers**

The IEE (Institution of Electrical Engineers) compiles the INSPEC database which contains records of the papers in much of the world's published scientific and technological literature dating back to 1969. It is the largest database in the physico-technical domain, holding over 5 million records of publications in the areas of computing, electrical engineering, electronics, physics and IT. Optoelectronics and optics is a major subject area in the INSPEC database. In 1991 the database contained some 190 000 abstracts on the two topics, with a further 1300 abstracts being added each month. One of the main advantages of using INSPEC lies in the fact that it covers a broad spectrum of publications, ranging from purely academic and scientific journals to the more product-oriented technological domain. INSPEC has been able to provide a good picture of the firms' overall optoelectronics-related activities.

An online search was conducted to extract the abstracts of papers written for journals and conferences by scientists working in the firms listed above between 1976–89, using a group of keywords related to optoelectronics. A meeting was first held with the Online Marketing officer of INSPEC to discuss the most effective search procedure. After studying the 'Search guide to Optics and Optoelectronics in the INSPEC Database' compiled by IEE, I chose a number of keywords in such a way as to cover the main areas of optoelectronics. These could be broken down into the following eight groups: Table 5.1 lists the categories in each group.

*Table 5.1* Grouping scheme of INSPEC data

Group A: Optical Communications Systems
Digital Communication Systems, LAN, Cable TV, ISDN
Subscriber loop, Optical Link, Optical
Communication, Module, Optical Communication
Equipment, Optical Information Processing,
Optical Interconnect, Optical Switching Systems

Group B: Optical Fibre Related
Optical Cable, Optical Fibre,
Optical Fibre Sensor, Optical Image Fibre, Optical Sensor

Group C: Optoelectronic Key Components
Semiconductor Laser, Distributed Feedback Laser,
Laser Mode, Laser Theory, Digital Simulation,
Light-Emitting Diode, Photodiode, Avalanche
Photodiode, Charge Coupled Devices (CCDs),
Integrated Optoelectronics, Integrated Optics,
Optical Waveguide, Optical Switch, Monolithic
Integrated Circuit

Group D: Underlying Generic Component Technologies
Chemical Vapour Deposition, Vapour Phase Epitaxy,
Liquid Phase Epitaxy, Molecular Beam Epitaxy,
Crystal Defects, Electron Device Noise,
Dislocation, Life Testing, Photoluminescence, TEM
SEM, Xray Examination

Group E: Semiconductor-Related Technologies
III-V Semiconductors, II-VI Semiconductors,
Amorphous Semiconductors, Epitaxial Growth, IC
Technology, Semiconductor Technology

*Table 5.1—continued*

| | |
|---|---|
| Group F: | Optical Disks<br>Video and Audio Disks, CD-ROM, Magneto-Optical<br>Device, Optical Disk Storage, Optical Storage,<br>Magnetic Thin Film, Magneto-optical Recording,<br>Thermomagnetic Recording, Optical Material |
| Group G: | Liquid Crystal Displays<br>Liquid Crystal Display, Liquid Crystal Device,<br>Thin Film Transistor, Nematic Liquid Crystal,<br>Electroluminescent Display, Light Emitting Device,<br>Thin Film Device |
| Group H: | Other Applications<br>Laser Beam Application, Laser Printer, Scanner,<br>Printed Circuit Board, Computer Vision, Pattern<br>Recognition, Picture Processing,<br>Electrophotography |

Group A includes optical communications systems such as long distance communications, LAN (local area networks) and optical interconnect systems. Group C includes the main optoelectronic components such as semiconductor lasers, distributed feedback lasers, light-emitting diodes (LEDs), photodiodes, CCDs (Charge Coupled devices) and OEICs (optoelectronic integrated circuits). The list in group D contains the generic technologies which have been used to develop the optoelectronic key components, such as epitaxial growth techniques. Group E is not directly linked to optoelectronics but was included since optoelectronics competence building is closely related to that of semiconductor and IC fabrication technologies.[10]

The above categorization scheme was designed by the author in order to cover the three largest industrial sectors which had been affected by optoelectronics: communications, IT and consumer electronics. The classification was designed to highlight the interaction between systems, products, key components and the generic component technologies. For example, the areas of strength and weakness in optical communication systems, products, key components and the generic component technologies should be highlighted by examining the figures in groups A to E. The research focuses on the particularly important area of component generic technologies rather than those for opto-based systems which may be different. The examples of the latter include the various transmission technologies and network management skills.

## Strengths and Weaknesses of Bibliometric Analysis

Having obtained the results from an INSPEC search,[11] a bibliometric analysis was carried out for each firm. A manual bibliometric analysis provides the necessary data to perform a detailed study of the R&D capabilities and the advances achieved in a particular technological domain within firms. The level of detail presented by a bibliometric analysis is such that it can provide highly sensitive information.[12] INSPEC provides the author's name and institutional affiliation, date, document type, language, abstract, classification codes, controlled terms (CT), and supplementary terms (ST) in each record (Appendix F contains an example of an INSPEC record). INSPEC assigns the controlled terms which are equivalent to the keywords for each paper, so that they give a relatively unbiased representation of the subject. The supplementary terms (ST), in contrast, are the keywords assigned by the authors, and are less suitable for a bibliometric analysis, since they may be more subjective. The controlled terms in each record were checked, and the most appropriate ones were selected by reading through the abstract. In bibliometric analysis, it is possible to count the keywords either as fractions or as whole numbers. It was decided to use fractional counting; for example, in the case of a paper for which three key words were chosen, each would count as 1/3. A maximum of six controlled terms were chosen for each record (sometimes INSPEC lists more than six). As the selection of controlled terms depended on the author's judgement (the author has a degree in physics), after the controlled terms were chosen a random sample of about twenty papers were shown to a postdoctoral researcher working on optoelectronics in the physics department at the University of Sussex to see if the keywords I chose were accurate, and it was confirmed that this was the case. For each firm, the variations over time (1976–89) of the paper counts in the eight groups were summarized. In total 2652 abstracts were read and analyzed for the 11 companies. The INSPEC data were further broken down in terms of the main systems, key components and technologies to facilitate measurement of competences described below. Group C in the classification listed in Table 5.1 has been broken down into such key components as semiconductor lasers and LEDs. The final categories are: Optical communication systems, Optical fibres, Semiconductor lasers, Light-Emitting Diodes, Photodiodes, CCDs, OEIC, Epitaxy, III-V semiconductors, Optical Disks, and Liquid Crystal Displays.

In this research, a computerized online keyword search was used to construct an optoelectronics publication database for the 11 firms. The resulting database depended on the keywords chosen. If additional or fewer keywords had been selected, it would have led to the selection of more or fewer abstracts. An important question in this exercise is how well the database generated by the chosen keywords reflects the domain studied. Ideally, the dataset should not contain too many irrelevant papers, yet it should contain an adequate level of data to reflect the subject area. In the second stage of the bibliometric anaysis, while the database was scanned manually by reading through the abstracts and selecting the keywords, any spurious abstracts were discarded. The process of carrying out a bibliometric analysis is labour intensive but relatively straightforward.

Some weaknesses of the bibliometric analysis method should be mentioned. As Hicks, Martin and Irvine (1986) have pointed out, bibliometric data may often be biased against non-English speaking countries such as Japan. However, as INSPEC also covers many Japanese journals and conference publications, this problem should have been largely overcome. Also, the propensity to publish papers may be different across firms. Some firms may reward their scientists according to the number of patents and papers they have generated; in other words, the scientists in different firms may be under different pressure to write papers. This should not be a problem in this particular study, since we are interested in the *relative* counts of publications within different fields of firms and not the absolute levels. The last point to note is that even within the same firm, the propensity to publish papers may differ depending on the type of work. The author's interview at one firm belonging to group 1 revealed the following situation: scientists are encouraged to give papers at conferences where the participants may include potential customers, such as NTT. They are able to use conferences as an opportunity to present their technical expertise in front of their customers. If on the other hand the participants are mainly competitors, as in the case of a conference on optical disks, then there would be less incentive to present their work. This may explain why there may be more papers on components compared to final systems or materials. This point may cause some distortion in the analysis. However, this situation is likely to be the same everywhere and should affect the data in a similar way for all firms.

## The US Patent Data

We discussed earlier how US patent data could be as revealing as scientific publications data in capturing the dynamics of competence building in R&D. In a field like textile machinery the US patent classification codes are localized, but this is not the case for a new field like optoelectronics. As a result, the patent codes had to be

*Table 5.2* List of Optoelectronics-Related Patent Codes

Optical fibres-related including Systems
350/96.1–96.28; 350/96.3–96.33 ; 65/2; 65/3.1–3.15; 65/4.1–4.3

Semiconductor lasers
437/129–130; 357/17; 372/43–50; 372/75;

Light-emitting diodes
437/23; 437/127

Photodiodes
357/13; 357/19; 357/30–31

CCDs, image sensors
357/24; 437/53; 358/213.11–213.14; 364/237.82–237.83; 364/237.85; 382/65; 250/216; 250/227; 250/578;

Epitaxial growth, III-V semiconductor related
437/83–85; 437/90–97; 437/101; 437/104–108; 437/117; 437/119–126; 437/131–133;

Semiconductor, IC technology
156/625–634; 156/643; 156/648–649; 156/654–657; 156/600–606; 156/609–614; 156/620.2–620.3;

Optical disks
369/13; 369/44-46; 369/48; 369/100; 369/109-112; 369/ 121-122; 369/275; 369/ 249.4; 369/249.6; 346/135.1; 350/375–378

Liquid crystal displays, electroluminescent displays
350/330; 350/332–351; 340/784; 340/765; 427/109;

EL–340/781–782; 340/785; 340/756–763; 340/766; 340/771; 340/805;340/ 794–795 identified manually by checking the list of all patents granted in the past for each firm in the US Assignees Index. Once all the codes related to optoelectronics had been identified they were narrowed down further by counting their frequency of occurrence. Codes which occurred infrequently were deleted and the ones which were retained were grouped into 10 sub-fields in an attempt to match the INSPEC grouping scheme. Table 5.2 lists the US patent codes which were selected in order to cover the same main areas of optoelectronics as with the INSPEC publication data.

After the classification codes had been classified into the 10 groups, the processing and tabulation of the US patent data was done by the courtesy of US Department of Commerce, Patent and Trademark Office. The patent codes and their one line description were then checked manually to correct for duplicates, since some patents appeared in more than two sub-groups. Furthermore, a considerable amount of sorting had to be carried out to reassign the patents into the same classification scheme used in the INSPEC analysis.

**Interview Data**

Having completed the bibliometric analysis, I carried out interviews with the firms' managers and scientists to obtain historical accounts of the firms' entry into optoelectronics and the evolution of their technological trajectories. One of the main purposes of the interviews was to obtain a self-assessment of the firms' competences in the various fields of optoelectronics and to evaluate the extent to which these are reflected by INSPEC and patent data. The firms' scientists were asked to give a self-assessment of their technological competence in the sub-fields on a five-point scale relative to other technological fields within the company.

When several researchers and R&D managers were present at the interview, they provided the assessment scores after discussing these with one another. In some cases, two sets of assessment scores were provided, one which was technological (R&D) and the other related to overall competences including marketing and production. In such cases, technological competence scores have been used in the statistical analysis. A point to note is that we are referring here to the firms' internal strengths and weaknesses, and not to the external ones which are measured in relation to competitors.[13] The reasons behind the strengths and weaknesses and how they relate to the firms' previous accomplishments were discussed. Interviews enabled an understanding to be gained of how the optoelectronics-related trajectories arose and of the competence-building process. Of the various factors that are influential in this respect, the chief among them can be classified as technology-led, market-driven, competitor-driven and organizational-driven factors. The duration of company visits ranged between four and 16 hours, and two to six people were interviewed at each company.[14] Appendix G.1 summarizes the structure of the interviews and the main questions that were addressed during the interviews. One point to stress here is that I was not just obtaining information from the companies. During the interviews, after I had obtained their assessment scores, I would show the results of my bibliometric analysis to the scientists and R&D managers. This proved to be rather novel for them, revealing their strength and weakness in a way the scientists and managers had not seen before. This helped to make the companies interested in my research and encouraged them to be cooperative.

## Intra-firm Technology Indices

The use of Intra-firm Technology Indices (IFTI) is a novel technique developed to measure the revealed strength of a firm in particular technological fields. They are based on a standard economic calculation of Revealed Technology Advantage (RTA) indices developed by Soete and Wyatt (1983). They measure the comparative advantage of technological strength within sectors of a particular country. The areas of strengths and weaknesses may be computed by combining the use of RTA indices of scientific publications and US patent data. Sometimes known as the 'activity index', as Grupp (1993) points out,

> The term activity index is somewhat misleading as the denominator of that indicator contains all patents of that country X and all countries' patents in that field of technology T. Therefore strong activities might result in a low 'activity' value if either there is strong activity in other fields than T or other countries than X have little activities in other fields but the one under consideration.

Since we are interested in the competence-building process which is internal to the firm, the use of RTA indices of the two types of data may provide the foundation for measuring competences. It was stressed earlier that the dynamic aspect of competence building will be examined. The use of RTA indices measured over time would allow such dynamics to be captured. In this case, the technique is applied to measure the strengths and weaknesses of a firm in different sub-fields of a technological domain using INSPEC and patent data. IFTI is used here to measure the internal strength of a firm within optoelectronic-related areas. IFTI is defined as the firm's share of publications (or patents) within a field divided by that firm's share of total publications (or patents). Values of greater than unity indicate areas of relative strength within a field, and values less than one areas of relative weakness. IFTI values have been calculated from the cumulative counts in the subfields using INSPEC (abbreviated to $IFTI_{INS}$) and US patent data (abbreviated to $IFTI_{Pat}$). These are referred to as cumulative IFTI, as opposed to periodic IFTI, which are based on shorter time periods (for example 1976–80).

However, since the IFTI calculated in this way range from infinity to zero, with one being the mean level, they lack symmetry.[15] The indices have therefore been converted to a range between +100 and –100 using a hyperbolic function based on a formula introduced by Grupp (1993), so that the modulated IFTI values are symmetrical about zero.

## 5.3   EMPIRICAL RESULTS – THE ANALYSIS OF COMPETENCES

### Analysis of Papers

Figure 5.2 shows the bibliometric data expressed as percentages of publications per firm, for Sony, Philips, Fujitsu and Siemens, summarized into three periods. In order to illustrate what observations can be derived from such an analysis, we shall focus on Fujitsu as an example. From the data, it can be seen that over a 13 year period, the firm's scientists produced a total of 409 publications. Of these, 8.0% have been published in group A (communication systems) areas; 2.3% in group B (optical fibres); 42.3% in group C (optoelectronic key components); 8.0% in group D (generic component technologies); 26.5% in group E (semiconductor-related technologies); 3.0% in group F (optical disks); 3.1% in group G (LCDs) and 6.8% in group H (other applications). The paper counts were further broken down into systems, products, key components and generic

*Figure* 5.2 A summary of INSPEC bibliometric (percentage) date for four firms

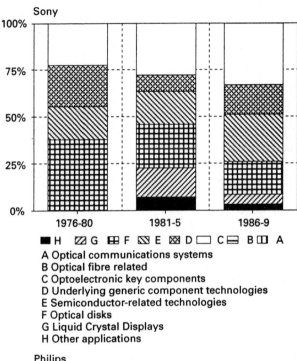

A Optical communications systems
B Optical fibre related
C Optoelectronic key components
D Underlying generic component technologies
E Semiconductor-related technologies
F Optical disks
G Liquid Crystal Displays
H Other applications

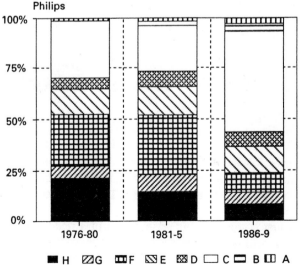

· *Figure 5.2—Continued*

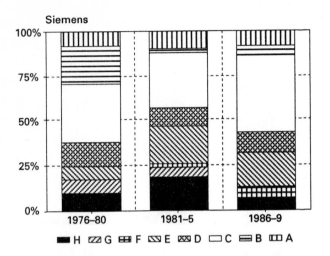

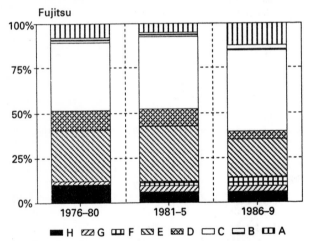

technologies for the 11 firms. Table 5.3 gives a breakdown of the papers into the various categories. 91 papers have been published in the area of semiconductor lasers and 108 in semiconductor-related technologies.

Even when it comes to the assimilation of a revolutionary technology like optoelectronics, the technological trajectories developed are likely to be influenced by the firm's core capabilities. The data highlight the emergence of similar trajectories at Sony and Philips, and at

*Table 5.3*  A summary of Fujitsu's INSPEC bibliometric data broken down into technological fields

|                     | 1976–81 | 1982–5 | 1986–9 | Sub-total |
|---------------------|---------|--------|--------|-----------|
| Communication System | 9.1    | 4.4    | 19.2   | 32.8      |
| Optical Fibre       | 3.7     | 1.8    | 3.9    | 9.5       |
| Semiconductor Laser | 21.3    | 26.2   | 43.7   | 91.2      |
| Light-Emitting Diode | 8.5    | 7.6    | 1.1    | 17.2      |
| Photodiode          | 8.0     | 11.2   | 11.7   | 31.0      |
| CCD                 | 0.0     | 1.3    | 1.5    | 2.8       |
| OEIC                | 3.7     | 13.3   | 14.5   | 31.5      |
| Epitaxy             | 10.8    | 15.5   | 6.5    | 32.8      |
| Semiconductors      | 29.5    | 45.8   | 32.9   | 108.2     |
| Optical Disk        | 2.3     | 2.0    | 8.0    | 12.3      |
| LCD                 | 2.0     | 5.5    | 5.0    | 12.6      |

CCD = Charge Coupled Device; OEIC = Optoelectronic Integrated Circuit; Epitaxy = Epitaxial Growth; LCD = Liquid Crystal Display.

Fujitsu and Siemens. These two pairs of companies belong to the two groups identified in Chapter 4. The common features of Sony and Philips are that both have a high concentration in optical disks (group F) and both are relatively weak in communications systems. The common features of Fujitsu and Siemens are that they have relatively high concentrations in optical communications (group A), and key components (group C), and relatively low concentrations in optical disks and LCDs (groups F and G). This is a reflection of their strengths in communications and weakness in consumer electronics. Fujitsu's high score for the semiconductor-related group (group E), is particularly noticeable. The data also reveal another point. The sub-totals of groups C, D and E account for over 60% of publications in all these firms (except for Philips where they account for 54%). This would suggest that investment in key components and component generic technologies is essential for building competences.

One significant trend at Sony has been the increase in key components and semiconductor-related technologies, reflecting Sony's attempts to complete a transition from being an AV (audio visual) maker to an AVCCC (audio visual, computer, components and communications) maker. This example suggests that competence building centres in areas closely related to the firm's core businesses. Once the competence-building process has reached a certain stage,

*Figure 5.3* Sumitomo Electric's bibliometric INSPEC data

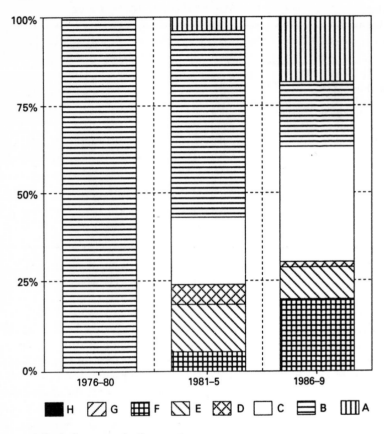

A Optical communications systems
B Optical fibre related
C Optoelectronic key components
D Underlying generic component technologies
E Semiconductor-related technologies
F Optical disks
G Liquid Crystal Displays
H Other applications

firms are likely to seize the opportunity to capitalize on economies of scope, and to branch into new areas.

Sumitomo Electric shows a different pattern from the other firms (see Figure 5.3). In 1976–81, the only area of activity was in optical fibres. During the 1980s, there was a trend to diversify into other areas of systems, key components and component generic technologies, in line with their VICC (Vertical Integration, Computer, Communications) strategy shown in Appendix E.3

The above examples demonstrate the use of bibliometric INSPEC data as a powerful tool for examining firm-specific trajectories. Technological strategies are reflected in the emerging trajectories, as in the case of Sumitomo Electric and Sony. They also highlight the areas in which firms have been concentrating their resources. $IFTI_{INS}$ per group for the 11 firms were calculated (see Table 5.4).[16]

*Table 5.4* Examples of time series $IFTI_{INS}$

*SONY*

|  | 1976–81 | 1982–5 | 1986–9 |
|---|---|---|---|
| Communication system | –100.0 | –100.0 | –99.5 |
| Optical fibre | –100.0 | –100.0 | –100.0 |
| Semiconductor laser | 9.5 | 27.8 | –2.0 |
| Light-emitting diode | –100.0 | –100.0 | –100.0 |
| Photodiode | –100.0 | –100.0 | –100.0 |
| Charge coupled device | –100.0 | 25.7 | 72.6 |
| OEIC | –100.0 | –100.0 | –97.6 |
| Epitaxial growth | 67.5 | –5.1 | 65.2 |
| Semiconductors | 5.8 | –22.0 | 27.8 |
| Optical disk | 92.3 | 73.5 | 67.3 |
| Liquid crystal display | –100.0 | –2.0 | –100.0 |

*STC*

|  | 1976–81 | 1982–5 | 1986–9 |
|---|---|---|---|
| Communication system | –9.4 | –56.1 | 39.0 |
| Optical fibre | –7.2 | 82.8 | 71.8 |
| Semiconductor laser | 39.0 | 55.2 | 41.2 |
| Light-emitting diode | –72.4 | –16.1 | –100.0 |
| Photodiode | –81.4 | 3.9 | –65.1 |
| Charge coupled device | –100.0 | –100.0 | –100.0 |
| OEIC | 56.3 | 30.5 | 9.5 |
| Epitaxial growth | 1.0 | 60.6 | 50.3 |
| Semiconductors | 33.1 | –53.6 | –13.8 |
| Optical disk | –100.0 | –100.0 | –100.0 |
| Liquid crystal display | 3.0 | –17.3 | –47.1 |

OEIC = Optoelectronic Integrated Circuit.

Up until this point, the IFTI figures shown have been based on cumulative counts of publications. In order to capture the dynamics of competence building, it is better to use time-based indicators. Table 5.4 presents the results of $IFTI_{INS}$ calculated over three time periods. In the case of Sony, we can see a rapid strengthening in CCDs. Communication systems has been an area in which Sony has been consistently weak. In the case of STC, we note that their areas of strength lie in optical fibres, semiconductor lasers and component generic technologies. On the other hand, they are relatively weak in LEDs, CCDs and optical disks, while their score in LCDs has been declining.

**Analysis of Patents**

The primary US patent data for Philips has been reproduced in Table 5.5. Philips was granted 671 optoelectronics-related patents during 1977–89. It has a total of 187 patents in the field of optical disks, followed by 108 in semiconductors. We also note a rapid increase in the number of CCD-related patents. Philips holds fewer patents in areas such as epitaxial growth, LCDs and components for communications.

*Table 5.5*   Summary of patent counts for Philips

|                             | 1977–81 | 1982–5 | 1986–9 | Sub-total |
| --------------------------- | ------- | ------ | ------ | --------- |
| Communication system        | 15      | 24     | 39     | 78        |
| Optical fibre               | 7       | 16     | 58     | 81        |
| Semiconductor laser         | 12      | 11     | 15     | 38        |
| Light-emitting diode        | 0       | 1      | 1      | 2         |
| Photodiode                  | 12      | 6      | 5      | 23        |
| CCD                         | 21      | 11     | 47     | 79        |
| OEIC                        | 0       | 1      | 1      | 2         |
| Epitaxy                     | 7       | 5      | 6      | 18        |
| Semiconductor               | 45      | 29     | 34     | 108       |
| Optical disk                | 54      | 67     | 66     | 187       |
| Liquid crystal display      | 11      | 8      | 22     | 41        |
| Electroluminescent-display  | 6       | 2      | 6      | 14        |
| Total                       | 190     | 181    | 300    | 671       |

CCD = Charge Coupled Device;
OEIC = Optoelectronic Integrated Circuit.

The patent counts of Fujitsu and NEC show that they are more specialized than the general electrical firms such as Siemens. The summary of the patents held by a firm in a technological field as a percentage of all patents in a field was also calculated. Cumulative $IFTI_{Pat}$ of the firms were evaluated using patent data. The patent counts and $IFTI_{Pat}$[17] were calculated by this grouping scheme, similar to the steps taken for INSPEC data.

*Table 5.6* Examples of $IFTI_{Pat}$ variation over time

| Sharp | 1977–81 | 1982–5 | 1986–9 |
|---|---|---|---|
| Communication system | –100.0 | –100.0 | –100.0 |
| Optical fibre | –100.0 | –100.0 | –99.0 |
| Semiconductor laser | –100.0 | –75.9 | 80.4 |
| Light-emitting diode | –100.0 | –100.0 | –68.8 |
| Photodiode | –70.0 | 7.7 | –22.0 |
| Charge coupled device | –100.0 | –93.0 | –57.4 |
| OEIC | 43.8 | –100.0 | –80.4 |
| Epitaxy | –100.0 | –81.5 | 62.5 |
| Semiconductor | –100.0 | –80.4 | –71.2 |
| Optical disk | –82.5 | –97.2 | –62.6 |
| Liquid crystal display | 84.2 | 89.6 | 71.0 |

| Siemens | 1977–81 | 1982–5 | 1986–9 |
|---|---|---|---|
| Communication system | 63.3 | 68.2 | 80.8 |
| Optical fibre | 66.3 | 83.6 | 55.5 |
| Semiconductor laser | –98.0 | –87.3 | –2.0 |
| Light-emitting diode | –66.3 | 75.4 | 88.9 |
| Photodiode | –54.9 | –13.8 | 20.4 |
| Charge coupled device | 17.2 | 3.0 | –63.8 |
| OEIC | 29.8 | 80.5 | 6.8 |
| Epitaxy | 27.1 | –66.3 | 0.0 |
| Semiconductor | –20.8 | –10.5 | –16.1 |
| Optical disk | –98.4 | –100.0 | –98.0 |
| Liquid crystal display | 30.5 | –26.8 | –93.0 |

OEIC = Optoelectronic Integrated Circuit.

Periodic IFTI$_{Pat}$ have been calculated for the 11 firms. Table 5.6 presents the data for Sharp and Siemens. Among the more significant features are Sharp's concentration in LCDs, its rapid advance in semiconductor lasers and component generic technologies, and its weakness in communications systems; Siemens in constrast has a high score in communication systems and LEDs, and average rating in semiconductors, and a low score in optical disks. These examples show how the use of periodic IFTI allows the dynamics of competence building to be captured.

**Interview Data**

Table 5.7 summarises the self-assessment values given by company scientists and managers. The values range from 5 (strongest) to 1 (weakest). The score may not be a whole number (as in the case of

*Table 5.7* Firms' self-assessment data for the sub-fields in optoelectronics

| | *Company* | | | | | | | | | |
| | *A* | *B* | *C* | *D* | *E* | *F* | *G* | *H* | *I* | *J* |
|---|---|---|---|---|---|---|---|---|---|---|
| CommSys | 3 | 4 | 1 | 0 | 5 | 4 | 4 | 5 | 3.5 | 5 |
| Optical Fibre | 1 | 5 | 0 | 0 | 3 | 0 | 4 | 0 | 0 | 0 |
| SemiLaser | 5 | 4 | 5 | 4.5 | 5 | 5 | 3.5 | 5 | 3.7 | 4 |
| LED | 4 | 4 | 5 | 5 | 0 | 5 | 5 | 3 | 5 | 5 |
| Photodio | 2 | 5 | 2 | 5 | 3 | 5 | 5 | 5 | 4 | 5 |
| CCD | 3 | 0 | 5 | 4 | 0 | 3 | 0 | 4 | 5 | 4 |
| OEIC | 4 | 5 | 3 | 4 | 4.4 | 4 | 4 | 3 | 3 | 5 |
| Epitaxy | 3 | 5 | 5 | 4 | 5 | 5 | 4.5 | 4 | 5 | 5 |
| Semicond. | 3 | 4 | 5 | 5 | 4 | 5 | 5 | 4 | 4 | 5 |
| Optical Disk | 5 | 0 | 5 | 5 | 0 | 0 | 2 | 3 | 4.5 | 3 |
| LCD | 4 | 1 | 1.5 | 5 | 3 | 5 | 2 | 2.5 | 4 | 2 |

*Source*: Author's interviews with the companies.
*Note*: Philips' data were not available.
CommSys = Communication Systems;
SemiLaser=Semiconductor Laser; LED=Light-
Emitting Diode; Photodio=Photodiode; CCD=Charge Coupled
Device; OEIC=Optoelectronic Integrated Circuit; Epitaxy-
Epitaxial Growth; Semicond.=Semiconductors;
LCD=Liquid Crystal Display.

semiconductor lasers) when the firm is engaged in several types of semiconductor lasers and an average has been calculated. Because of the confidential nature of self-assessment data, the company names will not be revealed here.[18]

## 5.4 TESTS OF CONSISTENCY OF THE VARIOUS MEASURES

### Cumulative Data

Having completed the analysis of papers and patents, calculated IFTI values based on these two sets of data, and obtained qualitative assessments through the interviews, we shall test the consistency of the various measures in this section. Once the different types of data are proved to be consistent with each other, we may use these results in section 5.4 to measure competences in firms. Table 5.8 gives a summary of the cumulative number of patents and papers related to optoelectronics produced by the firms between 1976 and 1989.

As can be seen from Table 5.8, there is considerable variation among the firms in their patenting which represents a partial indicator

*Table* 5.8   Cumulative total number of patents and papers

|          | *Patents* | *Papers* |
|----------|-----------|----------|
| NEC      | 212       | 460      |
| Fujitsu  | 191       | 409      |
| Sumitomo | 143       | 40       |
| GEC      | 46        | 113      |
| Siemens  | 553       | 213      |
| STC      | 53        | 128      |
| Toshiba  | 458       | 271      |
| Sharp    | 384       | 148      |
| Sony     | 206       | 91       |
| Philips  | 671       | 338      |
| Hitachi  | 688       | 441      |
| Total    | 3605      | 2652     |

of the firms' competences. In order to validate the consistency of the data, the empirical results of the tests have been summarized below. Table 5.9 presents the results from the various correlation tests between the cumulative data of patents and papers in the 11 fields for each firm. Later on we will conduct tests between the periodic data.

The tests are summarized as follows:

Test 1: Correlations between absolute cumulative counts of papers and patents

Test 2: Correlations between cumulative $IFTI_{INS}$ and $IFTI_{Pat}$

Test 3: Correlations between $IFTI_{INS}$ and self-assessment

Test 4: Correlations between $IFTI_{Pat}$ and self-assessment

Test 5: Correlations between absolute cumulative count of papers and self-assessment

Test 6: Correlations between absolute cumulative count of patents and self-assessment

*Table 5.9*  Results of the various tests on cumulative data

|  | *Test 1* | *Test 2* | *Test 3* | *Test 4* | *Test 5* | *Test 6* |
|---|---|---|---|---|---|---|
| Fujitsu | *** | * | ** | ** | X | X |
| Sony | ** | *** | *** | *** | *** | * |
| Sumitomo | *** | ** | * | ** | X | X |
| NEC | *** | ** | ** | *** | ** | ** |
| Toshiba | ** | ** | ** | * | * | X |
| Hitachi | *** | * | X | * | * | X |
| Sharp | *** | ** | ** | * | * | * |
| Siemens | * | ** | *** | * | ** | X |
| GEC | X | X | X | * | ** | X |
| Philips | *** | * | N. A. | N. A. | N. A. | N. A. |
| STC | ** | ** | *** | * | ** | X |

*Note*: 1-tailed test *** 1%, ** 5%, * 15% significance. N. A. Not available; X Not significant at <15%.

The results of Test 1 suggest that the correlations between cumulative counts of papers and patents are significant to a large extent. In 6 out of 11 cases the correlation coefficient is significant at the 1% level and 3 more at the 5% level. In 1 firm, the correlation coefficient between the cumulative $IFTI_{INS}$ and $IFTI_{Pat}$ is significant at the 1% level, in 6 more it is significant at the 5% level and 3 others at the 10% or 15% level. The areas of optoelectronics in which each firm has been concentrating as measured by scientific publications and patents are clearly related to one another. From the first two tests, however, we note that the data do not correlate for GEC. We may interpret this to be either a reflection of the weak linkage between research and product development in the company or the fact that at least one of the indicators may not be representing the full picture. The results highlight that, for the period under consideration as a whole, the relative areas of strengths as measured by IFTI ratios of patents and papers of the firms are consistent with one another.

Correlation tests have also been performed using the firms' self-assessments. If the data derived from INSPEC and US patents are meaningful that is to say, if the IFTI figures adequately represent the firms' strengths and weaknesses, we would expect them to correlate with the firms' self-assessment data, (which we assume are reasonably reliable). Table 5.9 summarizes the correlation tests between cumulative IFTI figures with the firms' self-assessments. In 3 out of 10 firms, the correlation coefficient between $IFTI_{INS}$ and the self-assessment is significant at the 1% level, a further 4 at the 5% level and 1 at the 15% level. For $IFTI_{Pat}$, the equivalent figures are 2 out of 10 at the 1% level, 2 more at the 5% level, 1 at the 10% level and the other 5 at the 15% level. Thus, although the degree of correlation is generally lower for patent-based data, in all 10 firms it is above the minimum significant level of 15%. The degree of correlation between the IFTI values and the self-assessments is similar for papers and patent data for some firms such as Fujitsu and Sony. The degree of correlation is extremely good for both patents and papers for Sony. This may suggest that Sony's competences are equally balanced between research and product development. However, for certain other firms, one type of IFTI correlates more closely with the self-assessment than the other. In the case of Sumitomo Electric, Hitachi and NEC, $IFTI_{Pat}$ correlate better, whilst for Toshiba, Sharp, Siemens and STC, $IFTI_{INS}$ correlate better. Let us examine possible reasons for this discrepancy.

**Reasons Why One Type of IFTI May be a Better Indicator Than the Other**

The first reason, which is supported by evidence from interviews with company management, relates to the particular circumstances of some firms concerning the procedures involved in filing for US patents. STC was owned by ITT until the mid-1980s and the filing for patents in the US had to be undertaken through its parent company ITT which continued to ignore requests from its subsidiary STC to file directly. There was no mechanism to check that the patents had been filed.[19] As a result, the patent profile for STC shows that it had no optoelectronics-related patents until the mid-1980s while it was owned by ITT, and shows signs of patenting in the field only after it was divested by ITT. GEC's management also stated that it was not in the company's interest to file for patents in the US.[20] For these two firms, this probably explains why INSPEC provides a better indicator than patents for assessing technological strengths and weaknesses.

A second possible reason is that, when firms are asked to assess their strengths and weaknesses in a certain technological domain, they are making an assessment based on their research capabilities as well as their technological capabilities. Since patent indicators lie towards the applied R&D end of the spectrum, they may not fully capture the firms' R&D capabilities in basic and applied research. INSPEC, on the other hand, covers a wider range of activities from the basic and applied end of research to product development, so it may be able better to capture the firms' competences.

A third possibility is that there may be a difference in the locus of the firms' strengths; in other words, some firms are more competent in their technological activities whilst others are more competent in their research capacities. Consequently one type of indicator will reflect the state of the firm better than the other. Examples of the former might include Sumitomo Electric, Hitachi and NEC, and of the latter Siemens and Toshiba. In section 5.2 we noted that there might be some distortion in the publication indicator, since the propensity to publish might depend on the type of work and therefore vary within the same organization. Such an error may be corrected by using data on patenting.

The above suggestions rest on the assumption that the firms' self-assessments are fairly reliable. The last possibility that might cause a discrepancy is that, although the IFTI figures give a fair snapshot of the firms' strengths and weaknesses, the self-assessment data may be skewed. This could be because the firms are not able to assess their strengths and weaknesses properly – in other words, they may not thoroughly understand their capabilities. Alternatively, some firms may have a distorted opinion of themselves, either being too harsh or too self-complacent. This might explain the case of Hitachi, where there is no correlation between the $IFTI_{INS}$ and the self-assessment, even though the correlation between the two sets of IFTI figures is statistically significant (as shown in Test 2 in Table 5.9). The self-assessment data may also be distorted because, when firms are asked to make quantitative assessments, they may judge themselves in terms of their performance during recent years rather than over a 13 year period. We will investigate this point later.

Further correlation tests were carried out with respect to the firms' self-assessment data. Tests 5 and 6 in Table 5.9 show the correlation coefficients between the absolute number of patents or papers and the firms' self-assessment. The correlation coefficients for the INSPEC data are significant at the 1% level for 1 firm, 4 firms at the 5% level, 1 firm at 10% level and 2 more at 15% level. The equivalent results for the patent data are just 1 firm at the 5% level, and 1 firm at the 10% level, and 1 more at the 15% level. Clearly, we have to rule out using absolute cumulative counts as reliable indicator of competences.

In short, a comparison of the above results suggests that the IFTI indicator is a better way of assessing a firms' strengths and weaknesses in different technological sub-fields.

## Correlation Tests Involving Periodic Data

Up to now, we have been checking the consistency of the cumulative measures over the whole 13 year period. As we noted earlier, it is quite possible that, when firms are asked to make a quantitative assessment, they may judge themselves in terms of their performance during recent times rather than the full 13 year period. Firms may even be assessing themselves with respect to changes in their competence over time. If they have been able to build their competence in a particular field relatively quickly and from a weak starting point, they might rate themselves highly. In order to add a time dimension to the analysis, I have carried out further correlation tests using periodic IFTI data.

The tests are summarized below:

Test 1: Correlations between periodic $IFTI_{INS}$ and $IFTI_{Pat}$ in 1986–9

Test 2: Correlations between periodic $IFTI_{INS}$ and $IFTI_{Pat}$ in 1982–5

Test 3: Correlations between $IFTI_{INS}$ in 1982–5 and $IFTI_{Pat}$ in 1986–9

Test 4: Correlations between $IFTI_{INS}$ in 1986–9 and self-assessments

Test 5: Correlations between $IFTI_{Pat}$ in 1986–9 and self-assessments

Test 6: Correlations between $IFTI_{INS}$ in 1982–5 and self-assessments

Test 7: Correlations between $IFTI_{Pat}$ in 1982–5 and self-assessments

One can observe from Test 5 in Table 5.10 that the correlation between $IFTI_{Pat}$ in 1986–9 and company self-assessments is significant in 8 out of 10 firms at the 10% level of significance. This compares with just 5 out of 10 in Table 5.9 (Test 4) using cumulative $IFTI_{Pat}$ (see above). This sheds some light on the way that firms

*Table* 5.10   Results of the correlation tests involving periodic data

|  | Test 1 | Test 2 | Test 3 | Test 4 | Test 5 | Test 6 | Test 7 |
|---|---|---|---|---|---|---|---|
| Fujitsu | * | ** | ** | X | ** | *** | * |
| Sony | ** | ** | *** | *** | * | ** | * |
| Sumitomo | * | * | *** | X | * | ** | * |
| NEC | X | *** | X | X | * | ** | ** |
| Toshiba | X | ** | * | * | ** | ** | *** |
| Hitachi | * | X | X | ** | * | X | X |
| Sharp | ** | * | * | ** | * | * | * |
| Siemens | *** | ** | *** | *** | *** | *** | X |
| GEC | * | X | * | * | X | * | X |
| Philips | X | X | * | N. A. | N. A. | N. A. | N. A. |
| STC | ** | * | ** | *** | * | ** | X |

*Note*: 1-tailed test *** 1%, ** 5%, * 10 or 15% significance. N. A. Not available.

may assess themselves by focusing on recent activities. A comparison of the results based on $IFTI_{INS}$ in Table 5.10 (Test 4) shows the slightly worse state of correlations, with 5 out of 10 firms significant at the 1% or 5% level, compared to 7 using cumulative $IFTI_{INS}$. However, of these, there is a strong degree of correlation for 3 firms, Siemens, Sony and STC. For the two firms, Hitachi and GEC, the results are better using time series $IFTI_{INS}$ in 1986–9. These firms may be judging themselves in terms of their recent state.

From the data in Table 5.9 we found that the correlation coefficients were significant at the 10% level or better between cumulative IFTI of papers and patent data in 9 out of 11 firms. When periodic IFTI are used, the number of firms drops to 6 in 1986–9 and 7 in 1982–5.

Test 3 in Table 5.10 shows the correlations between $IFTI_{Pat}$ in 1986–9 with $IFTI_{INS}$ in 1982–5. We mentioned earlier that in Japanese firms we might expect a time-lag of three years between INSPEC and US patent data due to the filing procedure. The results suggest that this is a valid assumption, as we find the correlation coefficients to be statistically significant in 7 out of 11 firms at the 10% level or better, with 2 more added at the 15% level. For some firms, namely Fujitsu, Sony, Sumitomo, GEC, Siemens, and STC, the correlation coefficients are better for these consecutive periods than for the entire 13 year period as shown in Table 5.9.

## 5.5  MEASURING FIRM COMPETENCES

The various tests have highlighted that there is no simple procedure for assessing firm competences. We have to examine the results of the tests for each firm and see whether any consistent picture emerges. That is to say, we have to make the best possible use of combining the cumulative IFTI, periodic IFTI, and self-assessment data. In doing so, the four best indicators will be selected on a case by case basis. This process is similar to focusing a camera; at first when one sees through the view finder the image may be blurred, but as one focuses it, one can obtain a clearer image.

The reasoning behind the selection process of the four sets of IFTI to be used for each firm will be discussed. We have summarized in Table 5.11, the results of the various tests in section 5.4 to check the validity of the different measures.

*Table 5.11*  Summary of the correlation tests for each firm

|        |     |      |      |      | Test |      |      |     |     |
|--------|-----|------|------|------|------|------|------|-----|-----|
| Firm   | 1   | 2    | 3    | 4    | 5    | 6    | 7    | 8   | 9   |
| Siemens | ** | *** | * | *** | *** | *** | X | *** | *** |
| Fujitsu | * | ** | ** | X | ** | *** | * | ** | * |
| NEC | ** | ** | *** | X | * | ** | ** | X | X |
| STC | ** | *** | * | *** | * | ** | X | ** | ** |
| GEC | X | x | X | * | X | * | X | * | * |
| Sumitomo | ** | * | ** | X | * | ** | .* | *** | * |
| Sony | *** | *** | *** | *** | * | ** | * | *** | ** |
| Philips | * | N. A. | N. A. | N. A. | N. A. | N. A. | N. A. | * | X |
| Sharp | ** | ** | * | ** | * | * | * | * | ** |
| Hitachi | * | X | * | ** | * | X | X | X | * |
| Toshiba | ** | ** | * | * | ** | ** | *** | * | X |

*Note*: 1-tailed test *** 1%, ** 5%, * 10 or 15% significance. x 20% significance; X does not correlate within 20% significance. N. A. not available.

Test 1: Cumulative $IFTI_{INS,}$ $IFTI_{Pat}$
Test 2: $IFTI_{INS}$, self-assessment
Test 3: $IFTI_{Pat}$, self-assessment
Test 4: $IFTI_{INS}(1986–9)$, self-assessment
Test 5: $IFTI_{Pat}(1986–9)$, self-assessment
Test 6: $IFTI_{INS}(1982–5)$, self-assessment
Test 7: $IFTI_{Pat}(1982–5)$, self-assessment
Test 8: $IFTI_{INS}(1982–5)$, $IFTI_{Pat}(1986–9)$
Test 9: $IFTI_{INS}(1986–9)$, $IFTI_{Pat}(1986–9)$

The selection criteria for the four sets of IFTI are based on the following: (1) which cumulative IFTI shows a high degree of correlation with self-assessment; (2) the degree of correlation between $IFTI_{INS}(1982–5)$ and $IFTI_{Pat}(1986–9)$, and that between $IFTI_{INS}(1986–9)$ and $IFTI_{Pat}(1986–9)$; (3) which periodic IFTI shows a high degree of correlation with self-assessment. Table 5.12 shows the four indicators chosen for Siemens. Table 5.13 lists the indicators used for all firms.

*Table* 5.12    Summary of Siemens' four indicators

| Cumulative $IFTI_{INS}$ | | $IFTI_{pat}$ (86–9) | | $IFTI_{INS}$ (82–5) | | $IFTI_{INS}$ (86–9) | |
|---|---|---|---|---|---|---|---|
| | | LED | 89 | LED | 90 | | |
| LED | 81 | CommSys | 81 | | | LED | 80 |
| CommSys | 66 | | | | | | |
| Opt. Fib. | 60 | | | | | | |
| | | Opt. Fib. | 56 | | | CommSys | 60 |
| | | | | Opt. Fib. | 41 | Photodio | 56 |
| | | | | Epitaxy | 38 | | |
| photodiode | 30 | | | | | | |
| Epitaxy | 30 | | | | | OEIC | 25 |
| OEIC | 24 | | | | | | |
| | | photodio | 20 | | | | |
| | | | | CommSys | 16 | | |
| | | | | SemiLaser | 14 | | |
| | | | | OEIC | 12 | | |
| | | | | Semicond. | 10 | | |
| | | OEIC | 7 | | | | |
| | | Epitaxy | 0 | | | Semicond. | –2 |
| Semicon | –3 | SemiLaser | –2 | | | Epitaxy | –2 |
| | | | | Opt. Disk | –9 | | |
| | | Semicon | –16 | | | | |
| SemiLaser | –27 | | | | | | |
| | | | | | | Opt. Fib. | –54 |
| Opt. Disk | –63 | CCD | –64 | | | LCD | –64 |
| | | | | | | SemiLaser | –77 |
| LCD | –88 | | | | | | |
| | | LCD | –93 | | | Opt. Disk | –94 |
| | | Opt. Disk | –98 | LCD | –100 | | |
| CCD | –100 | | | CCD | –100 | CCD | –100 |

## Standard Procedure

Having chosen the four sets of IFTI to use, a standard procedure was used to identify competences. For a given technological field, the weighted average was calculated according to the significance level of the correlations. For example, significance levels of 1%, 5%, 10% were given weights of 2, 1, 0.5 respectively. In the case of

*Table 5.13* The four sets of IFTI for each firm

| Firm | IFTI$_{INS}$ | IFTI$_{Pat}$ | Indicator IFTI$_{INS}$ (1986–9) | IFTI$_{Pat}$ (1986–9) | IFTI$_{INS}$ (1982–5) | IFTI$_{Pat}$ (1982–5). |
|---|---|---|---|---|---|---|
| Siemens | * | | * | * | * | |
| Fujitsu | * | * | | * | * | |
| NEC | * | * | | | * | * |
| STC | * | | * | * | * | |
| GEC | * | * | * | * | * | |
| Sumitomo | * | * | | | * | |
| Sony | * | | * | | * | |
| Philips | * | | | * | * | |
| Sharp | * | | * | * | * | |
| Toshiba | * | * | | * | | * |
| Hitachi | * | | * | * | | |

Philips, a simple average was used, since the IFTI figures could not be correlated with self-assessments. Once the competences had been calculated (they ranged between −100 and +100), they were linearly transformed to a value between 0 and 100 to facilitate mapping.

## Optoelectronics-related Competences of the Firms

Table 5.14 presents the results of this analysis.
Table 5.15 lists the areas of competence of the firms. They are the technological areas where the computed competences are greater than 60. Based on the areas of competence identified, it is possible to group the firms into two categories as shown below.

In some of the statistical tests in the following sections, it is necessary to aggregate the dataset into fewer technological fields. For the eleven firms, the technological fields were aggregated into five areas of:

1. Communications-related including optical fibres
2. Components for communications
3. Components for consumer and industrial electronics
4. Generic component technologies
5. Consumer products such as optical disks and LCDs.

The cumulative values of the competences in the five areas were evaluated. Table 5.16 gives a summary of the correlation matrix generated using those cumulative values. It represents the proximity of firms by their areas of strength and weakness. The analysis was done by SPSS. From Table 5.16 it can be seen that a strong positive correlation exists between Sumitomo Electric and GEC, STC and Siemens and STC and Sumitomo Electric, and also between NEC and Fujitsu, Fujitsu and Siemens and Sony and Toshiba. High correlations between Philips and Sony, and Philips and Toshiba are also noticeable. On the other hand, we find a high negative correlation between firms such as Sony and Sumitomo or STC, and between Toshiba and Sumitomo, GEC or STC and between Philips and Sumitomo, GEC, NEC or Siemens. Hence, the correlation matrix suggests that one can group the firms into communications-driven firms (namely Sumitomo, GEC, STC, Siemens, NEC and Fujitsu) and consumer-and industrial electronics-driven firms (Philips, Sony, Toshiba and Sharp). In the case of Hitachi, apart from a fairly strong correlation with Sharp, the correlation with other firms is relatively weak, which suggests that it is different from the others.

Table 5.14  Evaluated competences of the firms

| | Fujitsu | Sony | Sumitomo | NEC | Toshiba | Hitachi | Sharp | Siemens | GEC | Philips | STC |
|---|---|---|---|---|---|---|---|---|---|---|---|
| CommSys | 31 | 0 | 48 | 80 | 23 | 33 | 1 | 75 | 49 | 35 | 58 |
| Opt. Fib. | 13 | 0 | 99 | 21 | 4 | 18 | 6 | 63 | 67 | 41 | 87 |
| SemiLaser | 40 | 46 | 9 | 77 | 18 | 47 | 61 | 39 | 10 | 33 | 69 |
| LED | 41 | 22 | 29 | 29 | 80 | 34 | 6 | 93 | 32 | 17 | 14 |
| Photodio | 75 | 9 | 12 | 51 | 28 | 53 | 28 | 68 | 58 | 20 | 20 |
| CCD | 12 | 81 | 3 | 69 | 80 | 32 | 55 | 5 | 30 | 63 | 0 |
| OEIC | 72 | 1 | 73 | 22 | 11 | 61 | 19 | 59 | 87 | 15 | 54 |
| Epitaxy | 71 | 61 | 33 | 66 | 39 | 24 | 54 | 60 | 42 | 50 | 68 |
| Semicond | 76 | 47 | 29 | 35 | 68 | 46 | 51 | 49 | 36 | 39 | 37 |
| Opt. Disk | 16 | 87 | 16 | 27 | 60 | 60 | 59 | 17 | 1 | 79 | 2 |
| LCD | 11 | 26 | 0 | 24 | 39 | 77 | 78 | 7 | 60 | 38 | 37 |

CommSys = Communication System; Opt.Fib=Optical Fibre; SemiLaser=Semiconductor Laser; LED = Light-Emitting Diode; Photodio=Photodiode; CCD = Charge Coupled Device; OEIC = Optoelectronics Integrated Circuit; Epitaxy = Epitaxial Growth; Semicond=Semiconductors; Opt. Disk=Optical Disk; LCD = Liquid Crystal Display).

*Table 5.15*   Areas of strengths of the firms

| | |
|---|---|
| Siemens | Communication systems, Photodiodes, Epitaxy, LED, Optical fibres |
| Fujitsu | Photodiodes, OEIC, Epitaxy, Semiconductors |
| NEC | Communication systems, Semiconductor lasers, Epitaxy, CCD |
| STC | Optical fibres, Semiconductor lasers, Epitaxy |
| GEC | Optical fibres, OEIC, Liquid crystal displays |
| Sumitomo | Optical fibres, OEIC |
| Philips | Optical disks, CCD |
| Sony | Optical disks, CCD, Epitaxy |
| Sharp | Liquid crystal displays, Semiconductor lasers |
| Toshiba | Optical disks, Semiconductors, LED, CCD |
| Hitachi | Liquid crystal displays, Optical disks, OEIC |

OEIC = Optoelectronic Integrated Circuit.
CCD = Charge Coupled Device.

## 5.6   GROUPING FIRMS BY THEIR COMPETENCES

The final step in the analysis was to aggregate the competences into three fields for each firm. We can aggregate the competences of the technical fields identified into the following:

*Communications-related fields:*   Optical fibres, Comm. Systems, Photodiodes, OEIC
*Key components, generic technologies:*   Semiconductor laser, Epitaxy, Semiconductors
*Consumer/industrial electronics-related:*   LED, CCD, Optical Disks, LCD

In this way we should be able to see a clear pattern among the firms according to the loci of their strengths. Companies whose optoelectronics-related activities have centred on optical communications would be expected to score more highly in the first group of technical fields than the consumer/industrial electronics-related field. We would expect to see most firms having strength in generic component technologies such as epitaxial growth and semiconductors.

Table 5.16  Correlation matrix of the firms using competence scores

| | Sumitomo | Sony | Hitachi | NEC | Fujitsu | Toshiba | Sharp | GEC | STC | Philips | Siemens |
|---|---|---|---|---|---|---|---|---|---|---|---|
| Sumitomo | 1.0 | -.88** | -.10 | 0.55 | 0.21 | -.97** | -.72 | 0.81* | 0.89** | -.76* | 0.80* |
| Sony | -.88** | 1.0 | 0.10 | -.27 | 0.18 | 0.86** | 0.79* | -.45 | -.63 | 0.56 | -.55 |
| Hitachi | -.10 | 0.10 | 1.0 | 0.24 | 0.40 | -.10 | 0.69 | 0.03 | 0.16 | 0.29 | 0.06 |
| NEC | 0.55 | -.27 | 0.24 | 1.0 | 0.87** | -.47 | -.09 | 0.72 | 0.59 | -.83* | 0.94** |
| Fujitsu | 0.21 | 0.18 | 0.40 | 0.87** | 1.0 | -.18 | 0.33 | 0.63 | 0.45 | -.52 | 0.69 |
| Toshiba | -.97** | 0.86** | -.10 | -.47 | -.18 | 1.0 | 0.58 | -.80* | -.93** | 0.59 | -.71 |
| Sharp | -.72 | 0.79* | 0.69 | -.09 | 0.33 | 0.58 | 1.0 | -.34 | -.38 | 0.61 | -.40 |
| GEC | 0.81* | -.45 | 0.03 | 0.72 | 0.63 | -.80* | -.34 | 1.0 | 0.95** | -.71 | 0.83* |
| STC | 0.89** | -.63 | 0.16 | 0.59 | 0.45 | -.93** | -.38 | 0.95** | 1.0 | -.59 | 0.76* |
| Philips | -.76* | 0.56 | 0.29 | -.83* | -.52 | 0.59 | 0.61 | -.71 | -.59 | 1.0 | -.93** |
| Siemens | 0.80* | -.55 | 0.06 | 0.94** | 0.69 | -.71 | -.40 | 0.83* | 0.76* | -.93** | 1.0 |

*Note:* * indicates 2-tailed significance at 0.01 ;  ** at 0.001.

Thus, the second group should show high scores for both groups of firms. Companies whose optoelectronics-related activities have centred on consumer and industrial electronics applications would be expected to score higher in the third group of technical fields than the communications related field.

Table 5.17 lists the firms' aggregated competences. The firms are split into the three groups, according to the following criteria:

1. Communications-driven: The percentages of the communications related competences > 30% and the consumer/industrial electronics related competences < 30%.
2. Consumer/industrial electronics-driven: – The percentages of the consumer/industrial electronics-related competences > 30%, and the communications-related competences < 30%.
3. Evenly balanced: Others that are evenly balanced between communications and consumer/industrial electronics

*Table 5.17*  Aggregated competences of the firms

**1. Communications-driven**

| | Comm-related | Generic tech. | Consumer-related | Total |
|---|---|---|---|---|
| | % | % | % | |
| STC | 172(43) | 174(44) | 53(13) | 399 |
| GEC | 261(55) | 88(19) | 123(26) | 472 |
| Siemens | 265(48) | 165(30) | 122(22) | 552 |
| Fujitsu | 191(42) | 187(41) | 80(17) | 458 |
| NEC | 174(35) | 178(36) | 149(29) | 501 |
| Sumitomo | 232(66) | 71(20) | 48(14) | 351 |

**2. Consumer/industrial electronics-driven**

| | % | % | % | |
|---|---|---|---|---|
| Sharp | 54(12) | 166(40) | 199(49) | 419 |
| Sony | 10(3) | 154(41) | 216(57) | 380 |
| Philips | 111(25) | 122(28) | 204(47) | 437 |
| Toshiba | 66(15) | 125(28) | 258(57) | 449 |

**3. Evenly balanced**

| | % | % | % | |
|---|---|---|---|---|
| Hitachi | 164(37) | 117(23) | 206(39) | 487 |

Comm. related = Communications-related; Generic tech. = Generic technologies-related; Consumer-related = Consumer and industrial electronics-related.

*Figure* 5.4  Competence map of the firms

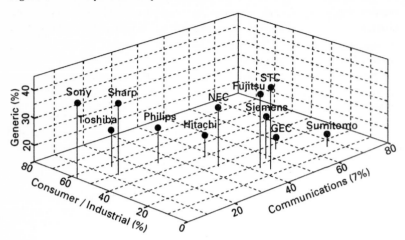

Figure 5.4 is a graphical presentation in three dimensions of the firms showing the percentages of the overall competences in communications, consumer/industrial electronics and component generic technologies fields. It suggests that there are two clusters of firms, one centred around Sharp, and the other around Siemens. Hitachi is located in between the two clusters. In other words, this figure shows graphically the grouping of the firms obtained in Table 5.17.

In short, the statistical analysis of competences has led to the same grouping of the firms into the three groups, which was derived earlier in Chapter 4 through historical, case-study analysis of the firms.

**Cluster Analysis**

Independent of the above analysis, cluster analysis was carried out of both the INSPEC and patent data using the SPSS statistical package. The data on paper (or patent) counts, IFTI and percentages of publications (or patents) for each firm was organized into five groups (optical communications, components for optical communications, components for consumer electronics, generic technologies and consumer or industrial electronics products such as optical disks and LCDs) and cluster analysis was carried out several times using various clustering methods. In the previous section we saw how the firms could be grouped into communications-related, or consumer and industrial electronics-related-driven firms. We would expect cluster analysis to result in a similar grouping. The results of the cluster analysis are shown below. The firms which are closely related to one another are listed in declining order of relatedness. In other words, the firms at the top of the list are most similar, followed by the next pair, and so on. A dendogram displays graphically the proximity of the firms to each other. A'+' indicates that the firms are related. The degree of proximity is indicated by the distance from the left – i.e. the proximity diminishes as one moves to the right.

*(a) Measure cosine method using INSPEC data*
With this method, NEC and Siemens are the most similar, followed by Sharp and Philips:

1. NEC, Siemens
2. Sharp, Philips
3. Sony, Toshiba
4. Sumitomo Electric, GEC
5. Fujitsu, STC
6. Sony, Sharp
7. NEC, Fujitsu
8. Sumitomo Electric, NEC
9. Sony, Hitachi

Figure 5.5 Dendogram based on measure cosine method using INSPEC data

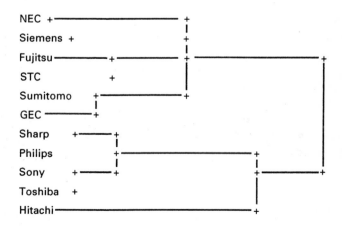

In the dendogram shown in Figure 5.5, we note that NEC and Siemens are closely related, and so are Sharp and Philips. Then there are two large group of firms consisting of those related to communications (NEC, Siemens, Fujitsu, STC, Sumitomo Electric and GEC) and those related to consumer electronics (Philips, Sharp, Sony, Toshiba). Hitachi is loosely connected to the latter group. In this way, cluster analysis has provided statistical support for the grouping of the firms into three categories derived in the previous section.

*(b) Centroid method using INSPEC data*
Using this alternative method, the clusters combined at each stage are shown below:

1. Sharp, Philips
2. NEC, Siemens
3. Sony, Toshiba
4. Hitachi, GEC
5. Sony, Sharp
6. Fujitsu, STC
7. NEC, Fujitsu
8. Hitachi, NEC
9. Sumitomo, Hitachi

With this method, the dendogram shown in Figure 5.6 was created. As in the above case, we can distinguish two main groups, one

*Figure* 5.6   Dendogram based on centroid method using INSPEC data

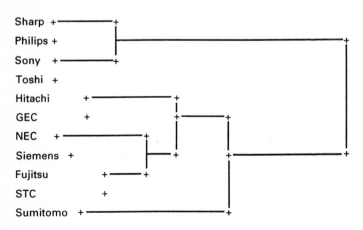

consisting of Sharp, Philips, Sony and Toshiba which are the consumer electronics-related firms, and the other consisting of Hitachi, GEC, NEC, Siemens, Fujitsu, STC the communications-related firms. Sumitomo Electric is an outlier loosely connected to the second group.

*(c) Centroid method using patent data*
Cluster analysis was run using patent data and the results of clustering at each stage and the dendogram are shown below and in *Figure* 5.7:

1. GEC, Siemens
2. Sumitomo, STC
3. GEC, Philips
4. Hitachi, Sharp
5. Sony, Toshiba
6. NEC, GEC
7. NEC, Fujitsu
8. Sony, NEC
9. Sony, Hitachi

Using patent data, we may observe from the dendogram the emergence of three groups consisting of (i) GEC, Siemens, Philips, NEC, Fujitsu – the communications-driven firms, (ii) Sony, Toshiba, Hitachi, Sharp – the consumer electronic-driven firms, and (iii) Sumitomo Electric and STC – the optical fibre-driven firms. Apart

Figure 5.7  Dendogram based on centroid method using patent data

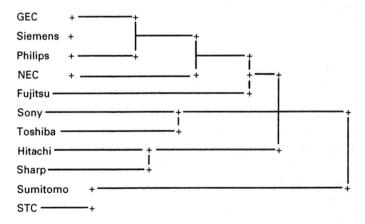

from clustering Philips into the communications-related group, the clustering seems to be consistent with the results using INSPEC data. The two groups are clearly distinguished from each other. In this section we have shown that we may categorize the firms into the three types by using a simple statistical clustering technique. We then arrive at similar results to those obtained in the previous section by assessing the competences of firms.

## 5.7  CONCLUSION

This chapter has presented the results of our statistical analysis of the competences of the firms between 1977 and 1989. The dataset consisted of 11 European and Japanese firms actively engaged in optoelectronics. We have seen how the use of bibliometric INSPEC data is a powerful technique for measuring technological trajectories. When firms have to assimilate a radical new technology such as optoelectronics, the assimilation pattern shows features that are strongly path dependent. For example, similar patterns were observed for Sony and Philips, and for Fujitsu and Siemens, reflecting similarities in their core businesses. The data also revealed that in most firms, the sub-totals of groups C, D, E account for over 60% of publications. Thus, investment in building capabilities at the level of key components and component generic technologies is essential for building competences. We shall present further evidence of path dependence in the next chapter, using other techniques.

The building blocks for formulating a technique for assessing technological competences have been developed using different types of data, through a process of trial and error. Statistical techniques were employed to test the validity of the three types of data: INSPEC bibliometric data, US patents and interview data. We found that the use of $IFTI_{INS}$ and $IFTI_{Pat}$ is a reliable way of assessing the strengths and weaknesses of firms in different technological fields, rather than other indices such as absolute counts and percentages of publications (or patents). We found that the use of IFTI figures coupled with the firms' self-assessment enabled us to gain an understanding of the areas of strength and weakness. However, for some firms the use of one type of IFTI is better than the other for particular reasons, while for other firms both types of IFTI are adequate indicators. For example, in the case of STC and GEC, $IFTI_{INS}$ was better than $IFTI_{Pat}$ since there are procedural reasons why US patents are not a useful indicator for these companies. We also discussed how periodic IFTI might provide a better picture than cumulative IFTI for certain firms. An interesting result here is the close correlation between $IFTI_{INS}$ for 1982–5 and $IFTI_{Pat}$ for 1986–9, confirming the postulated three year time lag between papers and patents. For certain firms, the correlations between IFTI and self-assessment are higher if one uses periodic IFTI rather than cumulative IFTI, reflecting the fact that there has been a rapid build up of competences in recent years. We also saw how one reason why the correlation between IFTI and self-assessments might be poor is because the self-assessment scores may be distorted. Several possible reasons for this were discussed. The various measures demonstrate that the areas of optoelectronics in which each firm has been concentrating (as measured by scientific publications and patents) are clearly related to one another. However, the various tests showed that it is not possible to have a simple automatic procedure for assessing the competences of firms. One has to look carefully at the various correlation tests to see whether a consistent picture emerges. The statistical analysis has shown that combination of $IFTI_{INS}$ and $IFTI_{Pat}$ coupled with company self-assessment enables an important, intangible phenomenon of competence building to be measured.

Having chosen four sets of IFTI to use, we employed a standard procedure to assess the competences of firms in 11 sub-areas of optoelectronics. These scores were then aggregated into three fields: communications-related, generic component-related technologies, and consumer and industrial electronics-related fields. Based on their scores in these three fields, the firms could then be categorized into three groups of: communications-driven, consumer and industrial electronics-driven; and evenly balanced firms. Firms in the first group are distinguished by their optoelectronics interests being driven primarily by optical communications. Firms in the second group are those whose optoelectronics-related activities centre on consumer and industrial electronics applications. Hitachi, the only firm in the third group, is a firm whose optoelectronics competence building has been evenly balanced between communications and consumer/industrial electronics. Cluster analysis of both the INSPEC and the patent data was carried out by SPSS using both INSPEC and patent data and resulted in a similar grouping of firms.

# 6 Learning, Interlinkages and Shifting Trajectories

The previous chapter presented the method and data employed to assess competences. We showed that the different types of data, namely US patents, INSPEC bibliometric data and self-assessments, are consistent with each another enabling technological competences to be measured. In Chapter 4 we examined how the path-dependent factors such as core businesses and top management strategy affected competence building. In this chapter, we set out to examine the role of organizational learning and interlinkages between component generic technologies, key components and systems. Although the final systems and products related to optoelectronics would be different among the firms reflecting their business interests, there would be common areas that most firms have been actively building capabilities in, centring on key components and the component generic technologies. Empirical evidence on the concept of interlinkages is presented, using two types of techniques, statistical analysis and technological linkage maps.

We also examine the concept of search zones and incremental learning. Firms make long-term investments in building technological capabilities which initially have few links with specific product applications. Competence building is characterized by trial and error and experimentation, which may eventually lead to fruitful outcomes. Firms search over a broader horizon in research initially and are able to gradually narrow down their search through a painstaking learning process. While competence building is constrained by path dependence, once competence building reaches a certain point, it provides the possibility for a firm to redirect its strategy into different paths.

## 6.1 SEARCH TRAJECTORIES

In this section, we explore the process of competence building as a long cumulative painstaking process, involving trial and error and considerable amount of learning. We will test the notion of search space by using Herfindahl[1] Indices. These indices serve to measure the technological diversification of firms at a given point in time. A value of one indicates concentration of activity in one field, and a value less than one indicates the broadening of activities. Table 6.1 summarizes the Herfindahl indices using INSPEC and patent data

*Table 6.1*   Herfindahl Index variation over time

*INSPEC*

| Company | 1977–81 | 1982–5 | 1986–9 | Average |
|---|---|---|---|---|
| Sumitomo | 1.0 | 0.36 | 0.26 | 0.54 |
| Sony | 0.36 | 0.32 | 0.28 | 0.32 |
| Hitachi | 0.26 | 0.25 | 0.27 | 0.26 |
| NEC | 0.31 | 0.24 | 0.23 | 0.26 |
| Fujitsu | 0.30 | 0.36 | 0.33 | 0.33 |
| Toshiba | 0.25 | 0.24 | 0.25 | 0.25 |
| Sharp | 0.33 | 0.28 | 0.31 | 0.31 |
| GEC | 0.36 | 0.24 | 0.31 | 0.30 |
| STC | 0.32 | 0.30 | 0.32 | 0.31 |
| Philips | 0.33 | 0.31 | 0.27 | 0.30 |
| Siemens | 0.23 | 0.25 | 0.29 | 0.26 |

*Patent*

| Company | 1977–81 | 1982–5 | 1986–9 | Average | H Index Higher* |
|---|---|---|---|---|---|
| Sumitomo | 0.76 | 0.63 | 0.66 | 0.68 | Yes |
| Sony | 0.36 | 0.44 | 0.44 | 0.41 | Yes |
| Hitachi | 0.24 | 0.29 | 0.31 | 0.28 | Yes |
| NEC | 0.25 | 0.20 | 0.22 | 0.22 | No |
| Fujitsu | 0.28 | 0.44 | 0.33 | 0.35 | Yes |
| Toshiba | 0.36 | 0.33 | 0.24 | 0.31 | Yes |
| Sharp | 0.90 | 0.68 | 0.37 | 0.65 | Yes |
| GEC | 0.34 | 0.47 | 0.21 | 0.34 | Yes |
| STC | 0.00 | 1.00 | 0.62 | 0.54 | Yes |
| Philips | 0.26 | 0.28 | 0.25 | 0.26 | No |
| Siemens | 0.25 | 0.28 | 0.34 | 0.29 | Yes |

H Index = Herfindahl Index.
*Average value of H index for patents is higher than that of INSPEC.

over three time periods for the 11 firms. As discussed earlier, we observe a diversifying trend for Sumitomo Electric, and a less obvious diversifying trend for Sony, NEC, Sharp, Philips based on INSPEC. On the other hand we note a focusing trend for Fujitsu and Siemens. Firms with the highest initial concentration in INSPEC seemed to have diversified most. On the other hand companies such as Hitachi and Toshiba have been diversified in optoelectronics research from the earlier period. These companies are more diversified than companies such as Sony or NEC, so we may be able to infer that the scope of competence building reflects the companies' overall scope of business activities. One important feature emerges from Table 6.1 relates to the concept of search zone. The average value of Herfindahl Indices is higher for patent data in 9 out of 11 firms. In other words, firms search over a broader range in basic and applied research while technology development is more focused and involves a narrower search zone.

Another type of quantitative evidence related to the reduction in search space coupled with learning is shown in Table 6.2. This table shows the correlations between IFTI over three time periods for both papers and patents. In 9 out of 10 firms, the correlation between $IFTI_{INS}$ in 1982–5 and 1986–9 is better than those between 1977–81 and 1982–5. In other words, the search space is becoming more focused in the later period compared to the earlier period. We may treat this result as more evidence of the way firms search over a broad horizon initially, and are able to gradually diminish their search space by a painstaking learning process. In the early phase of competence building, firms explore a broad range in order to deepen their understanding of the field. As they accumulate knowledge, they are able to narrow down the search path into areas which would strengthen their core businesses. With patent data, the result applies in 6 out of 11 firms. This is understandable, as activities of firms in patenting tend to be more focused, so that we do not expect to see such an improvement across firms. Nevertheless, the result supports the view that earlier periods characterised by uncertainty, trial and error do pave the way for activities which are better focused in later years.

Learning is an activity which takes time and effort and is not costless. Firms are able to reduce the search space by repeatedly performing various search routines and incremental learning. Suppose that learning was automatic and costless, and did not involve trial and

*Table* 6.2   Comparison of the correlations between the same type of IFTI
over three periods

IFTI$_{INS}$ CASE

| | 1977–81 with 1982–5 | | | 1982–5 with 1986–9 | | | |
|---|---|---|---|---|---|---|---|
| *Firm* | *Correl.* | *t-value* | *(%)* | *Correl.* | *t-value* | *(%)* | *Improved* |
| Fujitsu | 0.53 | 3.16 | 1% | 0.29 | 1.93 | 5% | No |
| Sony | 0.38 | 2.33 | 5% | 0.53 | 3.18 | 1% | Yes |
| Sumitomo | 0.45 | 2.69 | 5% | 0.52 | 3.13 | 1% | Yes |
| NEC | 0.36 | 2.24 | 5% | 0.44 | 2.64 | 5% | Yes |
| Toshiba | 0.12 | 1.09 | X | 0.79 | 5.88 | 1% | Yes |
| Hitachi | 0.00 | | X | 0.004 | | X | |
| Sharp | 0.03 | 0.50 | X | 0.49 | 2.96 | 1% | Yes |
| Siemens | 0.06 | | X | 0.44 | 2.67 | 5% | Yes |
| GEC | 0.01 | 0.31 | X | 0.10 | 1.01 | X | Yes |
| Philips | 0.15 | 1.26 | X | 0.31 | 2.01 | 5% | Yes |
| STC | 0.32 | 2.07 | 5% | 0.51 | 3.08 | 1% | Yes |

IFTI$_{Pat}$ CASE

| | 1977–81 with 1982–5 | | | 1982–5 with 1986–9 | | | |
|---|---|---|---|---|---|---|---|
| *Firm* | *Correl.* | *t-value* | *(%)* | *Correl.* | *t-value* | *(%)* | *Improved* |
| Fujitsu | 0.75 | 5.23 | 1% | 0.52 | 3.11 | 1% | No |
| Sony | 0.43 | 2.58 | 5% | 0.90 | 8.77 | 1% | Yes |
| Sumitomo | 0.14 | 1.23 | 15% | 0.39 | 2.39 | 5% | Yes |
| NEC | 0.21 | 1.54 | 10% | 0.06 | 0.74 | X | No |
| Toshiba | 0.31 | 2.02 | 5% | 0.57 | 3.49 | 1% | Yes |
| Hitachi | 0.61 | 3.73 | 1% | 0.62 | 3.80 | 1% | Yes |
| Sharp | 0.42 | 2.53 | 5% | 0.32 | 2.0 | 5% | No |
| Siemens | 0.33 | 2.12 | 5% | 0.46 | 2.77 | 1% | Yes |
| GEC | 0.34 | 2.16 | 5% | 0.14 | 1.23 | 15% | No |
| Philips | 0.32 | 2.07 | 5% | 0.15 | 1.24 | 15% | No |
| STC | 0.00 | | X | 0.45 | 2.72 | 5% | Yes |

error. Firms would be engaged in R&D and we would expect the fruit-ful outcomes of R&D to appear relatively early and stay constant. In other words we would expect the ratio of the types of research related to experimental, trial and error processes, applications of research, major breakthroughs, and those related to product development to remain more or less constant over the time horizon. If we allow the concept of learning and the reduction in search space to play some role, we might see a relative decline in the amount of research on ex-perimental trial and error type processes, and an increase in the R&D portion linked to product development over time to occur.

With INSPEC it is possible to do such an analysis, since the papers are categorised into different types of 'experimental', 'theoret-ical', 'practical/applications' and 'new development'. As the category suggests, 'experimental' papers are by nature describing experiments which serve to deepen the knowledge in the field but do not lead to major breakthroughs. 'Practical/applications' refer to papers which are related to practical applications, implying that the research has reached a stage where it can be applied to useful applications. 'New development'-type papers refer to cases where major breakthroughs have been achieved. In sum, if we count the types of papers over time in a given field and we observe a fall in 'experimental'-type papers and a rise in 'practical/applications' or 'new development'-type papers, we can draw the following conclusion. The competence-building process is a cumulative, painstaking, lengthy process, involving considerable difficulty. The initial long period entailed by uncertainty, trial and error and extensive experimentation may lead to successful fruitful outcomes linked to applications and new devel-opments in later years. Figure 6.1 shows the analysis of the types of papers over time for Philips, Sharp, Toshiba, Fujitsu, NEC, STC and Siemens, in the case of semiconductor lasers.

From Figure 6.1 a wedge shaped pattern can be observed, reflecting a decline in experimental papers and an increase in practi-cal applications and new developments for most firms, namely Sharp, Toshiba, Fujitsu, NEC and STC. For Toshiba for example, the percentages of experimental paper counts have fallen from 83.3% in 1976–81 to 52.6% in 1986–9 accompanied by an increase in the percentages of practical applications and new development-type papers from 16.7% to 32.9%. This suggests that the initial period of uncertainty, trial and error does lead to successful outcomes later. In addition to this conclusion, we may draw several other inferences from Figure 6.1.

The wedge shaped pattern is more pronounced for some firms than others. For example, in the case of Toshiba, NEC, and Fujitsu the pattern is more noticeable than say STC and Philips. The wedge pattern can hardly be seen for Siemens. We may be able to suggest from this analysis the emergence of a Japanese pattern, vis-a-vis the Europeans. One might conclude from this observation that Japanese firms are able to narrow down their search process better than the European firms. In other words, the Japanese firms may be better at organizational learning or at commercialization than their European counterparts. The wedge pattern is more pronounced for firms which have a high proportion of 'experimental'-type papers in the early period. The percentages of 'experimental'-type papers are higher for Japanese firms with the exception of Philips. A greater amount of experimentation in early years may pay off more in subsequent years.

The analysis in this section has shown that the concept of reduction in search space coupled with repeated trial and error and incremental learning process has a strong foundation.

*Figure* 6.1 Variation of the types of papers on semiconductor lasers over time

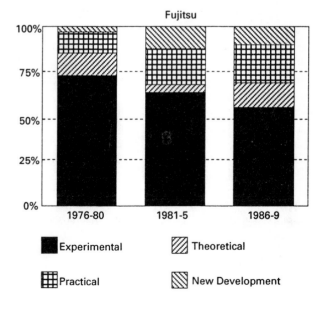

*Figure* 6.1—*Continued*

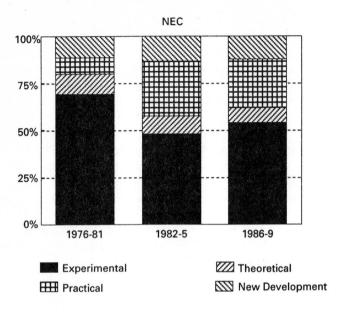

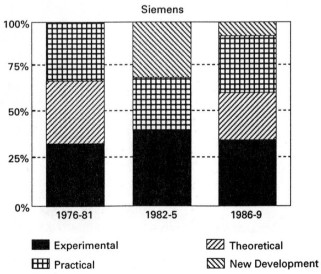

*Figure 6.1—Continued*

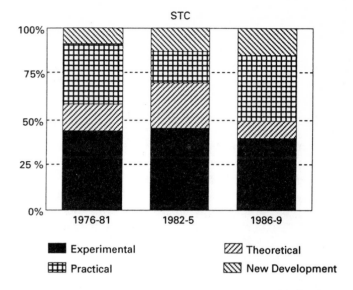

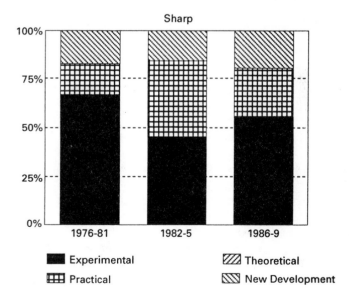

6.2  CONVERGENCE IN COMPETENCES

This section tests the relationship between systems, key components, and component generic technologies. Component generic technologies are the firms' underlying, enabling technologies that form the foundation for developing a wide range of key components. These key components are used as building blocks in a broad range of products and systems in different applications which are influenced by the firms' existing business and technological interests.

Suppose that the concept of the interlinkages between systems, key components and component generic technologies is not valid. Firms would search in areas of systems, products and component generic technologies in a random fashion. Measures of standard deviation of publications across firms would be more or less equal in all of these areas. Allow the concept of interlinkages to play some role, then we would expect variation across firms in systems and products to be large, reflecting the broadly different interests of firms by their business activities, and the variation across firms in some key components, and component generic technologies to be small, since they would be the common areas many firms would be actively exploring. Thus although the firms' final systems or product applications might be quite different, we shall show quantitatively that they are based on similar underlying component generic technologies.

If we allow the dimension of time to enter this model, we would expect the variance across firms to be larger in the earlier periods compared to the most recent period, as the nature of search would be more random and diverse to begin with but as more firms accumulate experience and their search pattern becomes more focused, the degree of variance would diminish across firms over time.

**Empirical Results Using Statistical Techniques**

Table 6.3 presents the results from a statistical analysis of the relative variance of the percentages of cumulative publications using INSPEC bibliometric data for the 11 technological fields of the 11 firms. For comparison, it also presents the mean value, standard deviation, minimum and maximum values. As expected, the coefficient of variation is the lowest for the group of technologies related to semiconductor fabrication, III–V semiconductors and other component generic technologies such as epitaxial growth. The values are respectively 27.3 and 37.4. The next lowest coeffcient of variation is

*Table 6.3*  Descriptive statistics of the percentages of cumulative publications of the 11 firms

|          | Mean   | StdDev | CoeffVar | Min.  | Max.   |
|----------|--------|--------|----------|-------|--------|
| Comm. Sys | 6.77%  | 4.56   | 67.3     | 0.2%  | 13.3%  |
| Opt.Fib  | 6.13   | 8.93   | 145.7    | 0.0   | 31.9   |
| Semilaser | 19.31  | 8.11   | 42.0     | 4.2   | 33.8   |
| LED      | 3.34   | 2.86   | 85.6     | 0.4   | 10.0   |
| Photodi  | 3.33   | 2.32   | 69.7     | 0.0   | 8.2    |
| CCD      | 4.52   | 4.31   | 95.4     | 0.0   | 11.4   |
| OEIC     | 5.15   | 4.23   | 82.1     | 0.5   | 13.8   |
| Epitaxy  | 8.88   | 3.32   | 37.4     | 2.7   | 13.7   |
| Semicon  | 18.98  | 5.18   | 27.3     | 10.5  | 28.5   |
| Opt.Disk | 9.26   | 7.94   | 85.7     | 0.5   | 23.9   |
| LCD      | 9.86   | 7.16   | 72.6     | 0.0   | 19.9   |

StdDev = Standard deviation; CoeffVar = Coefficient of Variance; Min. = Minimum; Max. = Maximum; Comm.Sys = Communication System; Opt.Fib = Optical Fibre; Semilaser = Semiconductor laser; LED = Light-Emitting Diode; Photodi = Photodiode; CCD = Charge Coupled Device; OEIC = Optoelectronic Integrated Circuit; Epitaxy = Epitaxial growth; Semicon = Semiconductors; Opt.Disk = Optical Disk; LCD = Liquid crystal display.

semiconductor lasers at 42.0, indicating that it is a key component firms in different final product markets have been working on. As expected, high values are measured for systems and products such as optical disks, LCDs optical fibres, communication systems at 85.7, 72.6, 145.7, 67.3 respectively.

The high coefficient measured for some components such as LEDs (light-emitting diodes), CCDs and OEICs indicates the greater spread of research activities across firms. One explanation is that LEDs are no longer key components but have become common components close to commodities. They have become easier to make, requiring no major innovations as a result of becoming mature and prices have continued to drop significantly. Some firms have already exited from the market. Application of CCDs is more limited than semiconductor lasers, as they are most commonly used in video cameras and photocopying equipment, thus limiting the scope of firms that are engaged in these areas.

*Table* 6.4 Descriptive statistics of the percentages of patents

|         | *Mean* | *StdDev* | *CoeffVar* | *Min.* | *Max.* |
|---------|--------|----------|------------|--------|--------|
| Group 1 | 24.4%  | 28.5     | 116.7      | 0.0%   | 78.3%  |
| Group 2 | 16.4   | 9.2      | 56.2       | 5.6    | 31.2   |
| Group 3 | 10.3   | 7.3      | 70.6       | 0.0    | 22.3   |
| Group 4 | 21.6   | 12.6     | 58.1       | 4.1    | 48.9   |
| Group 5 | 27.3   | 17.4     | 63.7       | 0.0    | 56.8   |

Table 6.4 relates to a similar analysis of the percentages of patents in the five groups of:

1. Communication systems including optical fibres
2. Key components for optical communications
3. Key components for consumer and industrial electronics
4. Generic component technologies
5. Consumer and industrial products (optical disks and LCDs

The effect is not as pronounced as for publications but is noticeable to some extent. The coefficient of variation is the highest for group 1 which is related to optical communications systems. Lower coefficients of variance are found in groups 2 and 4, which are the components for optical communications and the generic component technologies. However, the values are higher than for the INSPEC data, indicating a greater spread of activities across firms. This can be explained by the fact that technology development is a more focused activity than research, so that variances are higher in general.

Table 6.5 presents the results of the relative variance of cumulative $IFTI_{INS}$ based on INSPEC. The descriptive statistics are also summarized. As in the previous case, the lowest coefficient of variation is observed for the underlying component generic technologies such as semiconductors and epitaxial growth, at 29.7 and 37.9 respectively, followed by semiconductor lasers at 45.6. Relatively high values are measured for communication systems, optical fibres, optical disks and LCDs at 70.2, 73.9, 76.7 and 73.5. Higher values are also recorded for LEDs, photodiodes, OEICs and CCDs. The above results provide empirical evidence on the concept of generic, pervasive technologies that form the foundation of a variety of com-

*Table 6.5* Descriptive statistics of the cumulative IFTI$_{INS}$

|  | *Mean* | *StdDev* | *CoeffVar* | *Min.* | *Max.* |
|---|---|---|---|---|---|
| Comm. Sys | −13.1 | 61.0 | 70.2 | −99.8 | 57.6 |
| Opt.Fib | −10.3 | 66.3 | 73.9 | −100.0 | 97.2 |
| Semilaser | −12.1 | 40.1 | 45.6 | −92.3 | 45.0 |
| LED | −18.1 | 63.0 | 76.8 | −97.3 | 81.4 |
| Photodi | −21.6 | 53.9 | 68.3 | −100.0 | 64.5 |
| CCD | −22.3 | 73.4 | 94.1 | −100.0 | 70.4 |
| OEIC | −5.9 | 61.7 | 88.2 | −97.1 | 83.0 |
| Epitaxy | −1.9 | 37.2 | 37.9 | −81.7 | 43.9 |
| Semicon | −8.9 | 27.1 | 29.7 | −57.8 | 33.2 |
| Opt.Disk | −15.0 | 65.2 | 76.7 | −99.3 | 77.1 |
| LCD | −17.5 | 60.7 | 73.5 | −100.0 | 57.1 |

ponents. These technologies are of crucial importance to all firms engaged in optoelectronics. Also it has been shown that some of the key components such as semiconductor lasers are generic in nature, having a lower coefficient of variation. The generic component technologies and the key components form the building blocks of products and systems which are more diverse in nature, as reflected in the firms' cumulative business interests.

Table 6.6 relates to the descriptive statistics of the cumulative IFTI$_{Pat}$. As in the INSPEC case, lowest variances are observed for

*Table 6.6* Descriptive statistics of the cumulative IFTI$_{Pat}$

|  | *Mean* | *StdDev* | *CoeffVar* | *Min.* | *Max.* |
|---|---|---|---|---|---|
| Comm. Sys | −24.9 | 71.6 | 95.3 | −100.0 | 71.0 |
| Opt.Fib | −21.2 | 87.3 | 110.8 | −100.0 | 96.2 |
| Semilaser | −26.5 | 56.7 | 77.1 | −100.0 | 76.0 |
| LED | −25.7 | 73.6 | 99.1 | −100.0 | 92.2 |
| Photodi | −28.7 | 47.9 | 67.1 | −100.0 | 60.1 |
| CCD | −20.1 | 55.0 | 68.8 | −100.0 | 60.4 |
| OEIC | −42.4 | 70.9 | 123.1 | −100.0 | 81.5 |
| Epitaxy | −17.6 | 52.8 | 64.1 | −100.0 | 70.7 |
| Semicon | −16.7 | 51.2 | 61.5 | −91.84 | 61.4 |
| Opt.Disk | −36.6 | 66.5 | 104.9 | −100.0 | 76.3 |
| LCD | −24.2 | 61.4 | 81.0 | −100.0 | 82.0 |

the component generic technologies, such as epitaxy and semiconductors, followed by some key components, such as photodiodes, CCDs and semiconductor lasers. One can see higher variances in optical fibres, communication systems, optical disks, OEIC and LCDs. As in Table 6.4 we note higher variances compared to publications in most fields. Statistically, it was thus possible to show the existence of the interlinkages between products, components and the generic component technologies.

Table 6.7 summarizes the results of the coefficient of variation calculated over three time periods as measured by the percentages of INSPEC publications aggregating the 11 technological fields into the 5 groups. From Table 6.7 the coefficients of variation are the largest for the earliest period 1977–81 across the 5 groups, and diminish towards the recent period of 1986–9 as predicted. These results indicate that there is greater variation across firms in the earlier period in their activities due to a broader search space, whilst in the recent period, the difference between them has diminished, leading to a greater convergence of activities. Within each time frame, the lowest coefficient is observed in group 2, the key components including semiconductor lasers, and group 4, the generic component technologies.

Table 6.8 summarizes the results of the coefficient of variation calculated over three time periods as measured by $IFTI_{INS}$ of the 5 technological groups. The results are similar to the above case. The coefficient of variation is the largest for the earliest period and diminishes towards the more recent period indicating a greater convergence of competences. Within each time frame, the lowest coefficient is found in either group 2 or 4, the key components for communications or generic component technologies.

## Technological Linkage Maps

Technological linkage maps were produced based on INSPEC and interview data. The origins of technological linkage maps can be traced to STA (Strategic Technical Area)/product line matrix introduced by Mitchell (1986). The format of the maps was devised in order to express in a concise manner the linkage between component generic technologies, key components and systems. Since it is not possible to present the three levels on one map, it is split into two. The first type of map (TYPE I) shows the interlinkages between component generic technologies and the key components. The second type of map (TYPE II) shows the interlinkages between key components, other generic technologies and systems. The vertical column of

*Table 6.7* Descriptive statistics of the percentages of publications over three periods

| *1977–81* | *Mean* | *StdDev* | *CoeffVar* | *Min.* | *Max.* |
|---|---|---|---|---|---|
| Group 1 | 16.9 | 28.7 | 169.7 | 0.0 | 100.0 |
| Group 2 | 25.6 | 13.2 | 51.6 | 0.0 | 42.4 |
| Group 3 | 10.0 | 12.0 | 120.1 | 0.0 | 40.0 |
| Group 4 | 25.7 | 15.2 | 59.3 | 0.0 | 42.7 |
| Group 5 | 22.7 | 17.3 | 76.3 | 0.0 | 52.0 |

| *1982–5* | *Mean* | *StdDev* | *CoeffVar* | *Min.* | *Max* |
|---|---|---|---|---|---|
| Group 1 | 14.5 | 15.8 | 108.6 | 0.0 | 53.4 |
| Group 2 | 27.0 | 9.6 | 35.7 | 15.5 | 42.8 |
| Group 3 | 8.6 | 4.9 | 57.1 | 2.0 | 19.1 |
| Group 4 | 29.4 | 6.8 | 23.1 | 21.0 | 45.5 |
| Group 5 | 20.6 | 15.0 | 72.9 | 5.4 | 46.0 |

| *1986–9* | *Mean* | *StdDev* | *CoeffVar* | *Min.* | *Max* |
|---|---|---|---|---|---|
| Group 1 | 12.1 | 10.2 | 84.5 | 0.4 | 36.8 |
| Group 2 | 33.4 | 10.0 | 29.9 | 22.3 | 47.2 |
| Group 3 | 7.4 | 5.3 | 71.9 | 0.0 | 16.2 |
| Group 4 | 26.2 | 7.8 | 29.6 | 10.4 | 39.3 |
| Group 5 | 20.2 | 10.6 | 52.1 | 7.5 | 38.1 |

StdDev = Standard deviation; CoeffVar = Coefficient of Variance; Min. = Minimum; Max. = Maximum.

Group 1 = Communications systems including optical fibres.
Group 2 = Key components for optical communications.
Group 3 = Key components for consumer and industrial electronics.
Group 4 = Generic component technologies.
Group 5 = Consumer and industrial products (optical disks and LCDs).

Table 6.8 Descriptive statistics of the cumulative IFTI$_{INS}$ over three periods

1977–81

| | Mean | StdDev | CoeffVar | Min. | Max. |
|---|---|---|---|---|---|
| Group 1 | -19.1 | 66.7 | 82.4 | -100.0 | 97.8 |
| Group 2 | -26.1 | 41.7 | 56.5 | -100.0 | 28.4 |
| Group 3 | -15.0 | 76.7 | 90.2 | -100.0 | 93.6 |
| Group 4 | -20.7 | 48.7 | 61.4 | -100.0 | 30.1 |
| Group 5 | -6.1 | 67.0 | 80.9 | -100.0 | 75.9 |

1982–5

| | Mean | StdDev | CoeffVar | Min. | Max. |
|---|---|---|---|---|---|
| Group 1 | -7.7 | 75.0 | 100.5 | -100.0 | 92.7 |
| Group 2 | -8.8 | 32.1 | 43.6 | -53.1 | 39.8 |
| Group 3 | -14.5 | 48.2 | 74.1 | -90.8 | 63.3 |
| Group 4 | -9.4 | 20.7 | 38.3 | -39.1 | 34.7 |
| Group 5 | -17.9 | 61.6 | 75.0 | -87.5 | 66.0 |

1986–9

| | Mean | StdDev | CoeffVar | Min. | Max. |
|---|---|---|---|---|---|
| Group 1 | -11.2 | 61.6 | 69.3 | -99.8 | 83.3 |
| Group 2 | -3.7 | 28.0 | 29.1 | -40.1 | 31.3 |
| Group 3 | -16.9 | 59.2 | 71.2 | -100.0 | 62.8 |
| Group 4 | -2.5 | 30.3 | 31.1 | -72.1 | 39.4 |
| Group 5 | -12.9 | 46.2 | 53.1 | -77.2 | 53.3 |

Group 1 = Communications systems including optical fibres.
Group 2 = Key components for optical communications.
Group 3 = Key components for consumer and industrial electronics.
Group 4 = Generic component technologies.
Group 5 = Consumer and industrial products (optical disks and LCDs).

a TYPE I map lists the component generic technologies which are used for making the key components, listed horizontally, such as LEDs and semiconductor lasers. These in turn are the common building blocks for making the final products and systems. 'VI' in the map indicates 'Very Important', 'I' important.[2] Where possible, dates when the research started and ended are indicated. LCDs have been separated from other areas and special maps were made.

Comparing TYPE I maps across firms, there are many common features for all firms. In other words, if one looks at the maps say for Sony, Siemens, Toshiba and GEC (Tables 6.15(a), 6.12(a), 6.17(a), 6.13(a)) they are similar. Irrespective of the group to which the firm belongs, TYPE I maps are almost identical. These maps indicate the generic component technologies which are used to develop the key components. They are the common areas in which all firms have been building capabilities in, irrespective of the final product markets. Comparing TYPE II maps across firms, one notices differences among the firms. However, common features can be seen among the firms in the same group. For example the maps for NEC, Fujitsu, Siemens (Tables 6.9(b), 6.10(b), 6.12(b)) are similar and those of Toshiba, Sony and Sharp (Tables 6.17(b), 6.15(b), 6.16(b)) are also similar. Thus we can show through these interlinkage maps that the underlying component generic technologies which are common to all firms, enable key components to be made, which form the foundation for the firms to develop systems and applications which reflect their different business activities.

On examining TYPE I maps, one notes that some generic component technologies are more pervasive than others. For example, epitaxial growth techniques such as MOCVD or LPE are pervasive and very important for a range of components. Some others are of secondary importance such as TEM or SEM which are characterization techniques to measure the properties of the structures of crystals. One must note that some firms had been working on particular generic component technologies such as semiconductors and pulse code modulation for other applications before they entered the field of optoelectronics. Thus the firms have been building on their accumulated technological bases. One also notices the long time span over which some firms have been working on these technologies. Most firms have been working on epitaxial growth techniques for 10–20 years. It often takes a decade to develop capabilities in component generic technologies, and possibly another 10 years to reach commercialization.[3] In a way this is similar to how a pianist must continuously exercise scales every day in order to be good at playing masterpieces by Chopin or Beethoven or how an opera singer must continuously train her voice. On examining TYPE II maps, some key components are core in the sense that they are more pervasive than others, such as semiconductor lasers, while the maps also show that for some firms, the matrix is rather sparse (for example Table 6.13(b)). This may be an indication that the firm is not effectively capitalizing on economies of scope. On the other hand when matrices are dense (as in Tables 6.10(b), 6.18(b)), they may suggest that the firms are pursuing a policy of capitalizing on economies of scope. What TYPE II maps show is the scope of the firms' involvement in optoelectronics. Thus on looking at GEC's Type II map (Table 6.13(b)), six out of ten applications are related to defence.

In the case of firms in group 1, the maps show how continuous accumulation in generic component technologies enable firms to develop more technically advanced key components (such as moving from InGaAsP lasers to DFB lasers) over time. This strategy leads to development of higher performance systems, as reflected by the increased bit rate per second of optical communications, as in the case of NEC and Fujitsu. The case of Fujitsu should be given special mention. Although there had been pressure to put an end to the MBE (molecular beam epitaxy) project, the research continued for a long time and finally led to fruitful outcomes in several areas, including key components for optical communications such as semiconductor lasers, but also for another device called HEMT (High Electron Mobility Transistor) which is used in satellite communication receivers, for which the company achieved a world's best record in performance.[4] This example shows that investment in generic technologies enables fruitful outcomes in unforeseen areas.

In the case of liquid crystal displays (see Table 6.16(c)), it can be seen that some technologies are core in the sense that they are prerequisite for all generations of LCDs. Examples are LCD materials, polarizers, etching and microscopic connection technologies. These technologies have been developed over many years. Some technologies are applicable for only a limited type of applications. In any application, there are some technologies which are crucially important. In calculators, for example, they were driver circuits, control circuits, multiplexing and twisted nematic materials. The timing and arrival of CMOS technology was just right for LCDs. Kodama (1991) also discusses this point. In active type displays, the crucial

*Table 6.9a* NEC's TYPE I technological linkage map

| Component / Generic Technologies | Visible LED (1960s–1985) | LED for comm. (1976–84) | APD (Mid 1970s– Early 1980s) | PIN-PD (late 1970s) | AlGaAs Laser (1964–78) | Hi-Power AlGaAs Laser (1978–) | InGaAsP Laser for comm. (1976–) | Visible Wvlength Laser (1979–) | DFB Laser (1983–) | OEIC (1980–) | CCD (1981–) |
|---|---|---|---|---|---|---|---|---|---|---|---|
| LPE (1964–87) | VI | | | | VI | VI | VI | I | I | VI | I |
| MOCVD (Early 1970s–) | VI | VI | | | VI | I | VI | VI | VI | VI | I |
| VPE (1970s–) | | | VI | VI | | VI | VI | VI | | VI | VI |
| MBE (Early 1980s) | | I | | VI | VI | VI | VI | I | | I | I |
| Optical Mod (1964–) | I | I | | | I | VI | VI | I | VI | I | |
| Heterojn (1970s–) | I | I | VI | VI | VI | VI | VI | VI | VI | VI | |
| Degradation (1964–) | I | I | I | I | VI | I | VI | VI | VI | I | I |
| Life Testing (1964–) | VI | VI | VI | VI | 1964–9 | 1979–85 | 1978–83 | VI | 1983–8 | VI | I |
| Etching (1960s–) | I | I | I | I | VI | VI | I | I | I | I | |
| Photomask (1960s–) | I | I | I | I | I | I | I | I | I | VI | VI |
| Metalization (1960s–) | I | I | VI | I | VI | I | I | I | I | I | |
| Coating (1976–) | | | VI | | | VI | | | | I | |
| TEM (1960s–) | I | I | I | | I | I | I | I | I | I | |
| SEM (1960s–) | I | I | | | | | | | | I | |
| Photolum (1960s–) | I | I | | | I | I | I | I | I | I | |
| Xray Diff. (1960s–) | I | I | I | I | I | I | I | I | I | I | |

*Note:* for abbreviations in Tables 6.9a–6.18c, see pp.186–87.

Table 6.9b  NEC's TYPE II technological linkage map

| Systems — Key Components and Generic Technologies | CD Player CD-ROM (Mid 1980s–) | Laser Beam Printer, (1980–) | 34 M b/s Opt. comm. (1967–) | 100 M b/s Opt. comm. (Early 1970s–) | 280 M b/s Opt. comm. (1970s–) | 565 M b/s Opt. comm. (Late 1970s–) | 1.2 G b/s Opt. comm. (Early 1980s–) | 10 G b/s Optical Comm (1986–) | Coherent trans. (1988–) |
|---|---|---|---|---|---|---|---|---|---|
| LED (1960s–85) | | | VI | VI | | | | | |
| PIN-PD (Late 1970s–) | | | VI | I | VI | VI | VI | VI | VI |
| APD (Mid 1970s–Early 1980s) | | | | VI | VI | VI | VI | VI | VI |
| AlGaAsP Laser (1964–78) | VI | | VI | | | | | | |
| Hi Power | VI | | | | | | | | |
| AlGaAs Laser (1978–) | | | VI | VI | | | | VI | |
| InGaAsP Laser (1976–) | | | | | VI | VI | VI | VI | |
| Visible Wvlength Laser (1979–) | VI | VI | | | | | | | |
| DFB Laser (1983–) | | | | | | | VI | VI | VI |
| OEIC (1980–) | | VI | | | | VI | I | I | VI |
| CCD (1981–) | | VI | | | | | | | |
| IC circuit Tech. (1960–) | I | VI | I | VI | VI | VI | VI | VI | VI |
| Optical Repeaters | | | | VI | VI | VI | VI | VI | I |
| Digital comm. (1957–) | | | VI | VI | VI | VI | VI | VI | VI |
| PCM (1957–) | | | VI | VI | VI | VI | VI | VI | VI |
| WDM (Mid 1980s–) | | | | | | VI | VI | VI | VI |
| FDM (Mid 1980s–) | | | | | | VI | VI | VI | VI |
| ASK detection (Mid 1960s–) | | | VI | VI | VI | VI | VI | VI | VI |
| FSK detection (Mid 1960s–) | | | VI | | | VI | VI | VI | VI |

Table 6.10a  Fujitsu's Type 1 technological linkage map

| Component / Generic Technologies | LED for comm. (1973–82) | APD (1976–81) | PIN-PD (1978–84) | AlGaAs Laser (1976–83) | InGaAs Laser (1976–89) | DFB Laser (1983–) | OEIC (1979–) | HEMT (1978–89) |
|---|---|---|---|---|---|---|---|---|
| LPE (1970–) | I | VI | | VI | I | | | |
| MOCVD (1976–) | I | VI | VI | | VI | | VI | VI |
| MBE (1982–) | I | | | | I | VI | VI | VI |
| Opt. mod. (1970–) | I | | | | VI | VI | VI | VI |
| Heteroin (1970–) | VI | VI | VI | VI | VI | VI | I | VI |
| Crystal Growth (1960s–) | VI | I | I | VI | VI | VI | VI | VI |
| Degradation (1960s–) | I | VI | I | VI | VI | VI | VI | VI |
| Life Testing (1960s–) | VI | VI | VI | VI | VI | VI | VI | VI |
| Etching (Early 1960s–) | I | I | I | I | I | I | VI | I |
| Photomask (Early 1960s–) | I | | | | I | I | VI | I |
| Metallization (1960s–) | I | I | | | | | VI | |
| Coating (1970–) | | VI | | | VI | | | I |
| TEM (1974–) | I | I | | | I | | I | |
| SEM (1970s–) | I | I | | | I | | I | |
| Photolum. (Early 1960s–) | I | I | I | | | | I | |
| Xray Diff. (Early 60s–) | I | | | | | | I | |

*Table* 6.10b  Fujitsu's TYPE II  technological linkage map

| Key Components & Generic Technologies | CD-ROM. Opt. Stor. (Late 1980s–) | LBP (1980–) | 45 M b/s Opt. comm. (1972–81) | 100 M b/s Opt. comm. (1982–6) | 400 M b/s Opt. comm. (late 1970s) | 800 M b/s Opt. comm. (1980–) | 1.6 G b/s Opt. comm. (1984–) | 10 G b/s Opt. comm. (1986–) |
|---|---|---|---|---|---|---|---|---|
| LED for comm. (1973–82) | | | VI | I | I | | | |
| PIN–PD (1978–84) | I | | | VI | VI | VI | VI | VI |
| APD (1976–81) | VI | | VI | | | VI | | I |
| AlGaAs Laser (1976–83) | | | VI | | | | | |
| InGaAsP Laser (1976–89) | | | | VI | VI | VI | | |
| Visible Wvlength Laser (1983–) | VI | VI | | | | | | |
| DFB Laser (1983–) | | | | | VI | VI | VI | VI |
| OEIC (1979–89) | | VI | VI | VI | VI | VI | VI | VI |
| IC circuit Tech (1960s–) | I | VI | VI | VI | VI | VI | VI | VI |
| PCM (1950s–) | | | | | | | | |
| WDM (Mid 1980s–) | | | | | | VI | VI | VI |
| FDM (Mid 1980s–) | | | | | | VI | VI | VI |
| ASK detection (1983–) | | | | VI | VI | VI | VI | |
| FSK detection (1984–) | | | | VI | VI | VI | VI | |

*Table* 6.11a   Sumitomo Electric's TYPE I   technological linkage map

| Component Generic Technologies | LED (1975–) | PIN-PD (1978–) | Semiconductor Laser (1979–) | OEIC (1987–) | Optical Fibre (1967–) | Image Fibre (1980–) |
|---|---|---|---|---|---|---|
| LPE (1970–) | VI | | VI | VI | | |
| MOCVD (1980–) | | | VI | VI | | |
| Crystal Growth (1961–) | VI | VI | VI | VI | | |
| III-V Semicon. (1961–) | VI | VI | VI | VI | | |
| Opt. Mod. (1980s–) | I | | VI | VI | | |
| Heterojn (1975–) | I | I | VI | VI | | |
| Device Coating (1980s–) | VI | VI | VI | VI | | |
| Degradation (1970s–) | VI | VI | VI | VI | | |
| Life Testing (1970s–) | VI | VI | VI | VI | VI | VI |
| Reliability Testing (1967–) | VI | VI | VI | VI | VI | VI |
| Etching (1974–) | I | I | I | I | | |
| Photomask (1974–) | I | I | I | I | | |
| Metallization (1974–) | I | I | I | I | | |
| VAD (1975–80) | | | | | VI | VI |
| Cable Coating (1973–) | | | | | VI | VI |
| Computer sim. (Mid 1980s–) | | I | I | I | | |
| Large Diameter Drawing | | | | | I | I |
| Fibre Aging Tests (1968–) | | | | | VI | VI |

*Table* 6.11b   Sumitomo Electric's TYPE II technological linkage map

| Key Components & Generic Technologies | TV monitor Remote Observation System | LAN 300 Mb/s (1980–) | OPGW (1980s–) | Fiberscope, Opt. Sensor (1980–) | Infrared Opt. Fibre for Surgery |
|---|---|---|---|---|---|
| LED for Comm. (1975–) | VI | V | | | |
| PIN–PD (1978–) | VI | VI | | | |
| Semicon. Laser (1979–) | | VI | | | |
| OEIC (1987–) | | I | | | |
| Opt. Fibre | | | | | |
| Multimode | VI | VI | VI | VI | VI |
| Single Mode (1967–) | | VI | I | | |
| Image fibre (late 1980–) | VI | | | VI | |
| Cable connection (1950s–) | I | VI | VI | VI | VI |
| IC circuit Tech. (1970s–) | I | VI | VI | VI | I |
| PCM (1970s–) | | VI | I | | |
| Engineering Capability (1950s–) | | I | VI | VI | |

*Table 6.12a*  Siemens's TYPE I technological linkage map

| Component Generic Technologies | LED (1971–80) | LED for comm. (1981–87) | PIN–PD (Mid 1970s–) | AlGaAs Laser (1975–89) | InGaAsP Laser (1982–) | DFB Laser (Late 1980s–) | OEIC (1984–) |
|---|---|---|---|---|---|---|---|
| LPE (late 1960s–1991) | VI | VI | I | I | VI | I | VI |
| MOCVD (mid 1970s–) | | I | VI | VI | VI | VI | VI |
| MBE (late 1970s–) | | I | | VI | VI | VI | VI |
| Opt. mod. (1970s–) | | I | | I | VI | VI | VI |
| Heterojn (1960s–) | | I | I | I | I | I | I |
| Crystal Growth (mid 1950s–) | | I | VI | VI | VI | VI | VI |
| Degradation (1950s–) | | I | | VI | VI | VI | VI |
| Aging Test (1960s–) | | I | VI | VI | VI | VI | VI |
| Coating (1960s–) | | | I | I | I | I | VI |
| Etching (1950s–) | | I | I | I | I | I | VI |
| Photomask (1950s–) | | I | I | I | I | I | I |
| Metallisation (1950s–) | | I | I | I | I | I | |
| TEM (1950s–) | | I | I | I | I | I | I |
| SEM (1960s–) | | I | I | I | I | I | I |
| Photolum. (1960s–) | | I | I | I | I | I | I |
| Xray Diff. (1960s–) | | I | I | I | I | I | I |

*Table* 6.12b  Siemens' TYPE II technological linkage map

| Key Components & Generic Technologies \ Systems | Remote Control of Elec. eqpt. | LBP (1978–) | Cable TV Trans. (1977–) | 140 M b/s comm. Trans. for DPS (1970s–) | High Voltage Sensor | Comm. Netwk for Power Supply System (1982–) | 400 M b/s Opt. comm. (1982–7) | 2.4 G bit/s Opt. comm. (Mid 1980s–) |
|---|---|---|---|---|---|---|---|---|
| LED (1971–80) | VI | | | | | I | | |
| LED for comm. (1981–87) | | | VI | VI | | I | | |
| PIN-PD(Mid 1970s–) | | | I | I | | | VI | VI |
| AlGaAs Laser (1975–89) | | VI | | I | | | VI | I |
| InGaAsP Laser (1982–) | | | | I | VI | | VI | VI |
| DFB Laser (Late 1980s–) | | | | | | | I | VI |
| Laser beam Application (1960s–) | | VI | | | | I | | VI |
| OEIC (1984–) | | | | | | | I | I |
| Opt. Fibres (1970s–) | | | I | VI | I | I | I | I |
| Opt. Fibre Conn. (1970s–) | | I | I | VI | I | I | VI | VI |
| IC circuit (1950s–) | I | | I | VI | | I | VI | VI |
| System Design (1980s–) | | | I | VI | | I | | |
| Analogue Trans. | | | | | | | | |
| Digital Trans. | | | | VI | | | VI | I |
| PCM (1950s–) | | | | VI | | | VI | VI |
| WDM (Mid 1980s–) | | | | | | | | I |
| FDM (Mid 1980s–) | | | | | | | | I |
| ASK /FSK Detection (1980s–) | | | VI | | | VI | VI | |

Table 6.13a   GEC's TYPE I   technological linkage map

| Component Generic Technologies | LED (1950s–82) | PIN–PD (1985–) | Infra red PD (1986–) | GaAlAs Laser (Late 1960s–) | InGaAsP Laser (Late 1970s–) | OEIC (Mid 1980s–) | Bragg Cell (1983–) | TFT (1985–) | LCD (1982–) | SLM (1988–9) |
|---|---|---|---|---|---|---|---|---|---|---|
| LPE (1950s–) | I | I | I | I | | | | | | |
| MOCVD (Mid 1970s–) | I | I | I | VI | VI | VI | | I | | |
| MBE (Early 1980s–) | | | VI | I | VI | VI | | I | | VI |
| Epitaxial Growth (1960s–) | I | I | I | VI | VI | VI | | I | | |
| Opt. mod. (1960s–) | I | I | I | VI | VI | VI | VI | I | | |
| Heterojn (1960s–) | I | I | I | VI | VI | I | | I | | |
| III–V Semicon. (1962–) | I | I | I | I | I | I | | I | | |
| Degradation (1960s–) | I | I | I | I | I | I | | I | | |
| Aging Test (1960s–) | I | I | I | I | I | I | | I | I | |
| Computer sim. | | | | VI | I | | | | | |
| Coating | I | I | I | VI | I | VI | | I | | |
| Etching (1950s–) | I | I | I | I | I | VI | | I | | |
| Photomask (1950s–) | I | I | I | I | I | VI | | I | | |
| Metallisation (1950s–) | I | I | I | I | I | I | VI | I | | |
| Opt. Info. Proc. (1980s–) | I | I | I | I | I | | VI | I | I | VI |
| Semicon. Tech. (1950s–) | I | I | I | | I | | | I | VI | VI |
| LC Material (1982–) | | | | I | I | | | | VI | |
| TEM (1970s–) | I | I | I | I | | | | I | | |
| SEM (1960s–) | I | I | I | | | | | | | |
| Photolum. (1960s–) | | | I | | | | | | | |
| Xray Diff. (1960s–) | | | I | | | | | I | | |

*Table 6.13b* GEC's TYPE II technological linkage map

| Systems — Key Components & Generic Technologies | LED Display for Avionic (1980–2) | LCD Display for Avionic (1982–) | Infra red Image-Sensor (Late 1980s–) | Surveillance, Image Proc. | Radar (1980s–) | Process Control | Opt. Powered Sensor Netwk (1980s–) | Opt. Switch/ Opt. Interconnect (1980s–) | 565 M b/s Opt. comm. Netwk (Late 1970s–) |
|---|---|---|---|---|---|---|---|---|---|
| Visible LED (1950s–82) | VI | | | | | | | | |
| PIN-PD (1985–) | | | | | | | | | VI |
| Infra red PD (1986–) | | | VI | | | | | | |
| AlGaAs Laser (1960s–88) | | | | | I | | I | | I |
| InGaAsP Laser (Late 1970s–) | | | | | I | | I | | VI |
| DFB Laser (Mid 1980s–1989) | | | | | | | | | VI |
| OEIC (Mid 1980s–) | | | | | | | | I | VI |
| Bragg Cell (1983–) | | | VI | | VI | | | | |
| TFT (1980s–) | | VI | | | | | | | |
| LCD (1982–) | | VI | | | | | | | |
| SLM (1988–9) | | | | VI | | | | VI | I |
| Laser Beam Application | | | I | I | | | I | | |
| Semicon. Tech. (1950s–85) | | VI | I | VI | I | I | | | VI |
| DSP (1970–) | | | | I | VI | I | | | |
| Opt. Trans. | | | | I | | I | | | I |
| Opt. Fibre | | | | | | I | | I | VI |
| Opt. Fibre Sensor | | | | | | I | I | I | |
| Computer Tech. | | | | | | I | | | |
| Pattern Recognition (1983–) | | | | I | VI | | | | I |

Table 6.14a    STC's TYPE I  technological linkage map

| Key Components / Generic Technologies | LED Infra red (1962–70) | LED Visible (1966–75) | PIN-PD (1970s–) | AlGaAs Laser (1962–Mid 1970s) | Hi Power AlGaAs Laser (1986–87) | InGaAs Laser (Mid 1970s–) | DFB Laser (Early 1980s–) | OEIC (Mid 1980s–91) |
|---|---|---|---|---|---|---|---|---|
| LPE (Early 1960s–86) | I | I | VI | VI | | I | | I |
| MOCVD (Mid 1970s–) | I | I | VI | VI | VI | VI | VI | VI |
| Opt. mod. (1960s–) | I | I | | VI | VI | VI | VI | VI |
| Heterojn (Early 1960s–) | I | I | I | VI | I | I | VI | VI |
| III-V Semicon. (1960s–) | I | I | VI | I | I | I | I | I |
| Degradation (1960s–) | I | I | VI | VI | VI | VI (1980–7) | VI | VI |
| Aging Test (1960s–) | VI | VI | I | VI | I | VI | VI | VI |
| Coating (1960s–) | I | I | I | VI | I | I | I | I |
| Etching (Mid 1950s–) | I | I | I | I | I | I | VI | VI |
| Photomask (Mid 1950s–) | I | I | I | I | I | I | VI | VI |
| Metallisation (Mid 1950s–) | I | I | I | I | I | I | I | I |
| Computer sim. (Mid 1980s–) | | | | I | I | I | I | |
| TEM (1960s–) | I | I | I | I | I | I | I | |
| SEM (Early 1960s–) | I | I | I | I | I | I | I | |
| Photolum. (Early 1960s–) | I | I | I | I | I | I | I | I |
| Xray Diff. (Early 1960s–) | I | | I | I | I | I | I | I |

Table 6.14.b  STC'S TYPE II  technological linkage map

| Key Components & Generic Technologies | Range Finding for Military (1962–74) | 140 Mb/s, 100 Mb/s Opt. comm. (Late 1960s–70s) | 565 M bit/s Opt. comm. (1984–8) | LAN (Late 1970s–) | Cable TV Trans. (Late 1970s–Mid 1980s) | 2.4 G bit/s Opt. comm. (1986–90) | Coherent Trans. (1988–) |
|---|---|---|---|---|---|---|---|
| LED for comm. (1962–75) | I | | | | | | |
| PIN-PD (1970s–) | | I | VI | I | I | I | VI |
| AlGaAs Laser (1962–Mid 1970s) | VI | VI | | | | | |
| Hi Power AlGaAs Laser (1986–87) | | | | | VI | | |
| InGaAs Laser (Mid 1970s–) | | | VI | VI | | VI | I |
| DFB Laser (Early 1980s–) | | | VI | I | | I | VI |
| OEIC (Mid 1980s–1991) | | | | | | VI | I |
| Opt. Fibre (Late 1960s–) | | VI | VI | VI | VI | I | VI |
| Opt. Fibre Conn. (Late 1960s–) | | I | I | I | | VI | I |
| IC circuit (Mid 1950s–) | I | I | VI | I | VI | VI | VI |
| PCM, PIM (1960s–) | | VI | VI | | VI | | VI |
| WDM, FDM (Early 1980s–) | | I | I | | | | VI |
| ASK, FSK Detection (1980s–) | | | | | I | I | VI |
| Analogue Trans. (1930s–) | | | | VI | | | |
| Digital Trans. (1960s–) | | VI | VI | VI | | VI | VI |
| Network Design Mgt. (Mid 1980s–) | | | VI | | | VI | VI |

*Table* 6.15a  Sony's TYPE I technological linkage map

| Generic Technologies \ Key Components | LED (1970s–84) | AlGaAs Laser (1970s–84) | InGaAs Laser (1980–) | High Power Laser (1985–) | MQW Laser (1983–) | DFB Laser (1983–) | OEIC (1988–) | HEMT | CCD (1970–) |
|---|---|---|---|---|---|---|---|---|---|
| LPE (1970s–) | VI | I | | | | | I | | |
| MOCVD (1969–) | | VI | VI | | VI | VI | VI | VI | |
| MBE (1981–4, 1987–9) | I | VI | VI | VI | VI | VI | I | I | VI |
| Opt. mod. (1970s–) | I | VI | VI | VI | VI | VI | VI | VI | |
| Heterojn Interface (1970s–) | I | VI | VI | VI | VI | VI | VI | VI | VI |
| Degradation (1970s–) | I | VI | VI | VI | VI | VI | VI | VI | |
| Life Testing (1970s–) | I | VI | VI | VI | VI | I | VI | VI | VI |
| Etching (1960s–) | I | I | I | I | I | I | I | VI | |
| Photomask (1960s–) | I | I | I | I | I | I | | I | I |
| Metallization (1960s–) | I | I | VI | I | I | I | | I | VI |
| Coating (1970s–) | | I | | | | I | I | I | I |
| TEM (1970s–) | I | | | | | | I | | |
| SEM (1970s–) | I | I | I | | I | I | I | I | I |
| Photolum. (1960s–) | I | I | I | | | | I | I | |
| Xray Diffr. (1960s–) | I | I | I | | I | | I | | |

Table 6.15.b  Sony's TYPE II  technological linkage map

| Systems / Key Components & Generic Technologies | CD Player, VD Player (1974–) | Opt. Data Storage (1979–) | LBP OA eqpt (1980s–) | Welding, Medical Application (Mid 1980s–) | Other AV Eqpt (1980s–) |
|---|---|---|---|---|---|
| Visible LED (1970s–84) | | | I | | I |
| AlGaAs Laser (1970s–84) | VI | VI | | VI | VI |
| InGaAs Laser (1980–) | VI | VI | | | VI |
| Hi Power Laser (1985–) | | VI | VI | VI | |
| MQW Laser (1983–) | VI | VI | I | | I |
| DFB Laser (1983–) | | I | | | |
| OEIC (1988–) | I | I | | | |
| CCD (1970–) | I | | | | |
| IC circuit Tech. (1960s–) | VI | VI | VI | VI | VI |
| Signal Processing | VI | VI | VI | VI | VI |
| Opt. dis. media (1973–) | VI | VI | VI | | VI |

Table 6.16a  SHARP's TYPE I  technological linkage map

| Generic Component Technologies \ Key Components | Visible LED (Early 1960s–) | PD (1964–) | AlGaAs Laser (Mid 1970s–) | AlGaAs Laser (Visible & Hi Power) (1979–) | MQW Laser (1985–) | DFB Laser (1983–) | CCD (1978–) |
|---|---|---|---|---|---|---|---|
| LPE (1959–) | VI | VI | VI | I | | I | |
| MBE (1985–) | VI | | | I | VI | VI | |
| Optical mod. (1960s–) | I | VI | VI | VI | VI | VI | |
| Heterojn (1960s–) | | | VI | VI | VI | VI | |
| Laser Modes (1970–) | | I | VI | VI | VI | VI | |
| Degradation (1960–) | I | VI | VI | VI | VI | VI | VI |
| Life Testing (1963–) | VI | I | VI | I | VI | VI | VI |
| Etching (1960s–) | I | I | I | I | I | I | I |
| Photomask (1960s–) | I | I | I | I | I | I | I |
| Metallization (1960s–) | | I | I | VI | I | I | I |
| Coating (1960s–) | I | I | I | I | I | I | |
| Spectroscopy (1960s–) | | | I | I | I | I | I |
| SEM | I | I | I | I | I | I | |
| Photolum. | I | I | I | I | I | I | |
| Xray Diff. | | I | I | I | I | I | I |

Table 6.16b  SHARP's TYPE II technological linkage map

| Key Components & Generic Technologies | CD Player (Mid 1970s–) / VD Player (Late 1970s–) | Opt. Data Stor. (Mid 1970s–) | LBP, OA eqpt. (1970s–) | TV (1950s–) | HDTV (1985–) | Video camera, TV camera (Late 1970s–) |
|---|---|---|---|---|---|---|
| Visible LED (Early 1960s–) | I | | I | I | | |
| PD (1964–) | VI | I | | I | | |
| AlGaAs Laser (Mid 1970s–) | | VI | VI | | | |
| AlGaAs Laser (Visible and Hi Power) (1979–) | | VI | VI | | VI | |
| DFB Laser (1983–) | | VI | VI | | | |
| MQW Laser (1985–) | | VI | VI | | | |
| CCD (1978–) | | | VI | VI | VI | VI |
| LCD (1970–) | | | VI | VI | VI | I |
| IC circuit (1960–) | VI | VI | I | | VI | VI |
| DSP (1970s–) | VI | VI | | | | |
| Opt. dis. media (Mid 1970s–) | VI | VI | | | | |
| Precision Proc. (1970s–) | VI | VI | VI | | VI | VI |

Table 6.16c  SHARP's TYPE II technological linkage map for LCDs

| Systems | | SIMPLE MATRIX TYPE | | | | ACTIVE MATRIX TYPE | | |
|---|---|---|---|---|---|---|---|---|
| Generic Technologies | C O D E | Watches, Calculators (1970–9) | Word Processors with text (1979–82) | PCs, Text and Graphics (1982–9) | TV (1984–6) | PCs (1982–9) | TV (1984–9) | Projection Display (1985–9) |
| LCD Material | JD | | | | | | | |
| – DSM (1969–79) | | VI | | | | | | |
| – TN (1979–84) | | | VI | | | | | |
| – STN (1983–9) | | | | VI | VI | VI | VI | VI |
| – DSTN (1984–9) | | | | VI | VI | VI | VI | VI |
| Poralizer (1970–) | ID | VI | VI | VI | VI | VI | VI | VI |
| Spacer Technology (Late 1970s–) | ID | | I | I | I | VI | VI | VI |
| CMOS Driver (Late 1970s–) | ID | | | I | I | VI | VI | VI |
| Driver LSI (Early 1970s–) | ID | VI | VI | VI | VI | VI | VI | VI |
| Control circuit (Early 1970s–) | ID | | I | I | | I | I | |
| Interface circuit (1970s–89) | ID | I | I | I | I | I | I | I |
| Colour Filters (mid 1970s–89) | ID | | | VI | VI | VI | VI | VI |
| TFT array | ID | | | | | | | |
| – amorphous silicon (late 1970s–) | ID | | | | | VI | VI | VI |
| – Poly silicon (Early 1980s–) | ID | | | | | VI | VI | VI |
| Chip on Glass (Late 1970s–) | ID | | | | | VI | VI | VI |
| Surface Mounting (1970s–) | ID | | I | I | I | VI | VI | VI |
| Photomask (1960s–) | ID | I | VI | VI | VI | I | I | I |
| Etching (1960s–) | ID | I | I | I | I | VI | VI | VI |
| TAB (Early 1970s–) | ID | | VI | VI | VI | VI | VI | VI |
| Microscopic Conn. (1970–) | ID | I | VI | VI | VI | VI | VI | VI |
| Back Light (1970–) | JD | | VI | VI | VI | VI | VI | VI |
| Circuit sim. (1970s–) | ID | | | | | VI | VI | VI |
| Reliability Testing (1970s–) | ID | I | I | VI | VI | VI | VI | VI |
| Automatic Repair (1986–) | ID | | | VI | VI | VI | VI | VI |

Table 6.17a   Toshiba's TYPE I technological linkage map

| Component Generic Technologies | Visible LED (1970–) | LED for comm. (Early 1980s–) | APD (1982–4) | PIN-PD (1982–) | AlGaAs Laser (Mid 1970s–) | InGaAs Laser (1980–) | MQW Laser (1984–) | DFB Laser (1983–) | OEIC (Early 1980s–) | CCD (1970s–) |
|---|---|---|---|---|---|---|---|---|---|---|
| LPE (1970s–Late 1980s) | VI | I | I | VI | VI | I | | I | | |
| MOCVD (Early 1980s–) | | VI | I | VI | VI | VI | VI | VI | VI | |
| MBE (1984–) | | | | | | VI | VI | VI | VI | |
| Opt. mod. (1970s–) | I | I | VI | VI | I | VI | VI | VI | VI | |
| Heterojn (1970s–) | I | I | VI | I | VI | VI | VI | VI | VI | |
| Crystal Growth (1970s–) | VI | VI | I | VI | VI | VI | VI | VI | VI | VI |
| Degradation (1970s–) | I | I | VI | I | VI | VI | VI | VI | VI | I |
| Life Testing (1970s–) | VI | VI | I | I | I | VI | I | I | VI | I |
| Etching (1960s–) | I | I | I | I | I | I | I | I | I | I |
| Photomask (1960s–) | I | I | I | I | I | I | I | I | | I |
| Metallization (1960s–) | I | I | VI | I | I | I | I | I | | |
| Coating (1970s–) | | | | | VI | | I | I | | |
| Device Simulator (Mid 1980s–) | | | | | I | I | I | I | | |
| TEM (1970s–) | I | I | I | I | I | I | I | I | I | |
| SEM (1970s–) | I | I | I | I | I | I | I | I | I | |
| Photolum. (1960s–) | I | I | I | I | I | I | I | I | I | |
| Xray Diff. (1960s–) | I | I | | I | I | I | I | I | | |

*Table* 6.17b   Toshiba's TYPE II  technological linkage map

| Key Components & Generic Technologies | CD-ROM Opt. stor. (1970s–) | LBP. OA eqpt (1980s–) | Large Display Device | Video Trans. | LAN (1978–85) | 2.4 G bit/s Opt. comm. (1980s–) |
|---|---|---|---|---|---|---|
| Visible LED (1970–80) | | | VI | | | |
| LED for comm. (Early 1980s–88) | | I | | VI | | |
| APD (1982–4) | | | | | VI | VI |
| PIN–PD (1982–) | | | | | VI | VI |
| AlGaAs Laser (Mid 1970s–) | VI | VI | | VI | VI | VI |
| InGaAs Laser (1980–) | VI | VI | | VI | VI | VI |
| DFB Laser (1983–) | VI | | | VI | I | VI |
| MQW Laser (1984–) | | | | | I | VI |
| OEIC (Early 1980s–) | | I | | | | VI |
| CCD (1970s–) | | VI | | | | |
| IC circuit (1960s–) | I | VI | VI | VI | VI | VI |
| DSP (1970–) | VI | VI | I | VI | VI | VI |
| Optical Repeaters | | | | I | I | I |
| Analogue Trans. | | | | VI | | |
| PCM (1970s–) | | | | VI | VI | VI |
| WDM (1980s–) | | | | | | VI |
| FDM (1980s–) | | | | | | VI |
| ASK Detection (1980s–) | | | | | VI | VI |
| FSK Detection (1980s–) | | | | | VI | VI |

Table 6.17c  Toshiba's TYPE II  technological linkage map for LCDs

| Systems — Generic Technologies | SIMPLE MATRIX TYPE DISPLAYS | | | | ACTIVE MATRIX TYPE DISPLAYS | | | |
| --- | --- | --- | --- | --- | --- | --- | --- | --- |
| | Watches Calculators (1970s–79) | PCs with text (1979–82) | PCs with text and graphics (1982–7) | TV (1980–6) | PCs (1983–9) | TV (1984–9) | Passenger Aircraft Cockpit | Projection Display (1985–9) |
| LCD Material | | | | | | | | |
| DSM(1970s–79) | VI | | | | | | | |
| TN (1976–84) | | VI | VI | VI | | VI | VI | VI |
| STN(1984– | | | VI | VI | VI | VI | | VI |
| Ferroelectric (1986–) | VI | | | | | | | |
| Poralizer (1970s–) | VI | VI | VI | VI | VI | VI | VI | VI |
| CMOS Driver (1979–) | | VI | VI | VI | VI | VI | VI | VI |
| Driver LSI 1970s–) | VI | VI | VI | VI | VI | VI | | VI |
| Control circuit (1977–) | | I | I | I | VI | VI | VI | VI |
| Interface circuit (1970s–) | I | I | I | I | VI | I | VI | I |
| Colour Filters (mid 1970s–) | | | VI | VI | VI | VI | VI | VI |
| TFT array | | | | | | | | |
| – amorphous silicon (1983–) | | | | | VI | VI | VI | VI |
| – Poly silicon (1984–) | | | | | VI | VI | VI | VI |
| Chip on Glass (1980s–) | | I | I | I | VI | VI | VI | VI |
| Surface Mounting (1970s–) | | VI | VI | VI | VI | VI | VI | VI |
| Photomask (1960s–) | I | I | I | I | I | I | I | I |
| Etching (1960s–) | I | VI | VI | VI | VI | VI | VI | VI |
| TAB (1976–) | | VI | VI | VI | VI | VI | VI | VI |
| Microscopic Conn. (1970s–) | I | VI | VI | VI | VI | VI | VI | VI |
| Back Light (1976–) | | | | | VI | VI | VI | VI |
| Circuit sim. (1985–9) | | | | | | VI | VI | VI |
| Automatic Repair (1987–) | | | | | | | VI | VI |

*Table 6.18a* Hitachi's TYPE I Technological Linkage Map

| Key Components / Generic Technologies | Visible LED (1967–70) | LED for Comm. (1973–9) | APD (Early 1970s–) | PIN-PD (Early 1970s–) | AlGaAs Laser (1980–) | InGaAs Laser (1969–80) | MQW Laser (1984–) | DFB Laser (1975–) | OEIC (Mid 1980s) | CCD (1970s–) |
|---|---|---|---|---|---|---|---|---|---|---|
| LPE (1969–88) | VI | VI | | | VI | I | VI | VI | VI | |
| MOCVD (Early 1970s–) | | | I | I | VI | VI | VI | VI | VI | I |
| MBE (Mid 1980s–) | VI | VI | | | VI | VI | VI | VI | VI | |
| Opt. mod. (1969–) | VI | | VI | VI | VI | VI | VI | VI | VI | |
| Heteroin (1969–) | I | I | VI | VI | VI | VI | VI | VI | I | I |
| Crystal Growth (1960s–) | I | I | VI | VI | VI | VI | VI | VI | I | I |
| Degradation (1969–) | VI | I | I | I | VI | VI | VI | VI | VI | I |
| Life Testing (1969–) | VI | VI | VI | VI | VI | VI | VI | VI | VI | I |
| Etching (1960s–) | I | I | I | I | I | I | I | I | I | VI |
| Photomask (1960s–) | I | I | I | I | I | I | I | I | VI | VI |
| Metallization (1960s–) | I | I | I | I | VI | I | I | I | I | VI |
| Coating (1969–) | I | | | | VI | I | I | I | | |
| Device Simulator (Mid 1980s–) | I | | | | I | | | | | |
| SEM (1970s–) | I | I | I | I | I | I | I | I | | |
| TEM (1970s–) | I | I | | | I | I | I | I | I | |
| Photolum. (1960s–) | I | | I | I | I | I | I | I | I | |
| Xray Diff. (1960s–) | I | | I | I | I | I | I | I | I | |

*Table 6.18b* Hitachi's TYPE II technological linkage map

| Key Components & Generic Technologies | TV Trans. (1979–) | Nuclear Power Plant Netwk (1985–6) | Video Trans. (1979–82) | Computer Netwk 100 Mb/s (1980s–) | 2.4 G b/s Opt. comm. (Late 1980s–) | Coherent Opt. comm. (1988–) | OA Eqpt LBP (1980–) | Opt. stor. CD-ROM (1980s–) | CD player VD Player Video camera (1977–86) |
|---|---|---|---|---|---|---|---|---|---|
| Visible LED (1967–70) | VI | | I | | | | VI | | |
| LED for comm. (1973–9) | VI | VI | | VI | | | | | I |
| APD (Early 1970s–) | VI | VI | I | | VI | | | I | I |
| PIN-PD (Early 1970s–) | | | | | VI | VI | | | |
| AlGaAs Laser (1969–80) | | | VI | VI | VI | | I | VI | VI |
| InGaAs Laser (1980–) | | | | VI | VI | | VI | VI | VI |
| DFB Laser (1975–) | | | | | I | VI | VI | | |
| MQW Laser (1984–) | | | | | I | | I | | |
| OEIC (Mid 1980s–) | | | | | VI | VI | VI | | |
| CCD (1970s–) | | | | | | | | I | VI |
| Opt. dis. media (1970s–) | | | | | | | | VI | VI |
| IC circuit (1960s–) | VI | I | VI | VI | VI | VI | VI | VI | I |
| DSP (1970–) | I | VI | I | VI | VI | I | VI | VI | VI |
| Opt. Repeaters (1970s–) | I | I | I | I | I | I | | | I |
| Analogue Transmission | VI | VI | VI | VI | VI | VI | | VI | |
| Digital Transmission | | | | VI | VI | | | VI | |
| PCM (1970s–) | | | | | | VI | | | |
| WDM (Late 1980s–) | | | | | | VI | | | |
| FDM (Late 1980s–) | | | | | | VI | | | |
| ASK Detection (1980s–) | | | | VI | VI | VI | | VI | |
| FSK Detection (1980s–) | | | | VI | VI | VI | | | |

Table 6.18c   Hitachi's TYPE II technological linkage map for LCDs

| Systems | SIMPLE MATRIX TYPE DISPLAYS | | | | ACTIVE MATRIX TYPE DISPLAYS | | | |
|---|---|---|---|---|---|---|---|---|
| Generic Technologies | Watches Automobiles, Calculators (1970–9) | PCs with text (1979–81) | PCs with text and graphics (1982–5) | TV (1980–3) | PCs (1984–) | TV (1984–) | Projection Display (1986–) | Light Shutter Array (1985–) |
| LCD Material | | | | | | | | |
| DSM (1976–9) | VI | | | | | | | |
| TN (1978–84) | VI | VI | | | | | | |
| STN (1984–9) | | | VI | VI | | VI | VI | |
| Ferroelectric (1985–) | | | | | VI | | | VI |
| Poralizer (1976–) | VI | VI | VI | | VI | VI | | VI |
| CMOS Driver (1978–) | VI | | VI | VI | VI | VI | VI | VI |
| Driver LSI (1976–) | VI | VI | | | | | | |
| Control Circuit (1976–) | | I | VI | I | VI | VI | VI | VI |
| Interface Circuit (1976–) | I | I | VI | I | VI | VI | I | I |
| Colour Filters (mid 1970s–89) | | | VI | VI | VI | VI | VI | |
| TFT array | | | | | | | | |
| – amorphous silicon (1983–) | | | I | | VI | VI | VI | |
| – Poly silicon (1982) | | | | | VI | | | |
| Chip on Glass (1982–) | | | | | VI | | | |
| Surface mounting (1970s–) | | | | VI | VI | VI | VI | |
| Photomask (1960s–) | I | I | VI | VI | VI | VI | VI | |
| Etching (1960s–) | I | I | VI | VI | VI | VI | I | |
| TAB (1976–) | | I | VI | VI | VI | VI | VI | |
| Microscopic Conn. (1970s–) | I | VI | VI | VI | VI | VI | VI | VI |
| Back Light (1978–) | | | VI | VI | VI | | VI | |
| Reliabity tests (1970s–) | I | I | VI | VI | VI | VI | VI | |
| Automatic Repair (1986–) | | | | | VI | VI | VI | VI |

Note :

| | | |
|---|---|---|
| III–V semicon. | = | III–V semiconductors |
| APD | = | Avalanche photodiode |
| ASK | = | Amplitude Shift Keying |
| AV | = | Audio Visual |
| CCD | = | Charge Coupled Device |
| CD | = | Compact disk |
| Comm. | = | Communication |
| Computer sim. | = | Computer simulation |
| Conn. | = | Connection |
| DFB | = | Distributed Feedback |
| DPS | = | Data Processing System |
| DSM | = | Direct Scattering Mode |
| DSP | = | Digital Signal Processing |
| Elec. eqpt | = | Electronic equipment |
| FDM | = | Frequency Division Multiplexing |
| FSK | = | Frequency Shift Keying |
| HEMT | = | High Electron Mobility Transfer |
| Heterojn | = | Heterojunction Interface |
| HDTV | = | High Definition Television |
| Hi | = | High |
| Hi Power | = | High Power |

| | | |
|---|---|---|
| I | = | Important |
| IC | = | Integrated Circuit |
| ID | = | Internal development |
| Info. | = | Information |
| JD | = | Joint development with supplier |
| LAN | = | Local Area Network |
| LBP | = | Laser Beam Printer |
| LC | = | Liquid Crystal |
| LCD | = | Liquid Crystal Display |
| LED | = | Light Emitting Diode |
| LPE | = | Liquid Phase Epitaxy |
| MBE | = | Molecular Beam Epitaxy |
| Mgt. | = | Management |
| MOCVD | = | Metal Organic Chemical Vapour Deposition |
| MQW | = | Multi Quantum Well |
| Netwk | = | Network |
| OA | = | Office Automation |
| OEIC | = | Optoelectronic Integrated Circuit |
| Opt. dis. media | = | Optical disk media |
| Opt. Fibre | = | Optical Fibre |
| Opt. info. proc. | = | Optical information processing |

| | | |
|---|---|---|
| Opt. mod. | = | Optical modulation |
| Opt. stor. | = | Optical data storage |
| PCM | = | Pulse Code Modulation |
| PD | = | Photodiode |
| Photolum. | = | Photoluminescence |
| PIM | = | Pulse Interval Modulation |
| PIN–PD | = | PIN photodiode |
| Proc. | = | Processing |
| Semicon. Laser | = | Semiconductor Laser |
| SLM | = | Spatial Light Modulator |
| STN | = | Super-twisted Nematic (LCD) |
| TAB | = | Tape Automated Bonding |
| TEM | = | Transmission Electron Microscopy |
| TFT | = | Thin Film Transistor |
| TN | = | Twisted Nematic (LCD) |
| Trans. | = | Transmission |
| VAD | = | Vapour Phase Axial Deposition |
| VD | = | Videodisk |
| VPE | = | Vapour Phase Epoxy |
| VI | = | Very Important |
| Xray Diff. | = | Xray Diffraction |
| Wvlngth | = | Wavelength |
| WDM | = | Wavelength Division Multiplexing |

technologies are TFT arrays and tape automated bonding. Thus firms which succeeded in LCDs were able to leverage their competences in the underlying generic technologies as in CMOS. At Toshiba, strength in CMOS technology also enabled them to be competent in CCDs. Some technologies have become mature while others are still in the process of being developed. Examples of the former are driver circuits, polarizers and microscopic connection technologies; while examples of the latter are chip on glass, circuit simulation, TFT arrays and automatic repair technologies. The above discussion, based on the use of technological interlinkage maps, suggests that the model described earlier has a strong foundation. Firms invest in building capabilities at the level of component generic technologies over a long period.

## 6.3 THE ROLE OF KEY COMPONENTS

In some firms, technological investments are made with a long-term view to building technological capabilities, and considerations of short-term financial returns are of secondary importance. Having a solid foundation in component generic technologies allows them to develop key components, which in conjunction with other generic technologies provides the foundation to develop a broad range of value added systems.

There are a number of main reasons for developing key components in-house. Firms need key components with special characteristics to develop high value added products and systems. A key component is similar to the brain in a human being, as the firm cannot rely on rival makers to make key components. The purchase of key components on the open market from a supplier who is likely to be a competitor means that, at best, the key component would be the same as the ones being used by the supplier, but in general it is likely to be inferior. Furthermore, the supplier who might be a competitor would be able to infer the type of system the firm is trying to develop from the key components ordered. Thus in-house development would help enable maintenance of secrecy. There is the additional danger that the supplier may withhold the key components at some point in time.[5] In-house development of key components would also speed up the lead time to systems completion. Knowing the performance parameters of the key components enables the systems team to continue with their work while the key components are being developed, which leads to a reduction in overall lead time. Japanese firms have adopted this strategy. Some European firms such as

Philips and Siemens have also adopted this strategy even when their component businesses were making losses since losing the capabilities to develop key components affects product development at the downstream end.[6] Some European firms do not regard this as important; they will source key components from other makers if they are available on the market, and if it is less costly to do so. They might argue that given high fixed costs and low marginal costs, suppliers of components are very unlikely to withdraw supply. However, there is another reason in-house development is important. If one has to rely on part of the system from an external supplier, it would mean that one does not have full control over the whole system. In some cases the missing part might cause a bottleneck and adversely affect the development of the whole system. Lack of interchangeability leads to increased vulnerability.[7] In the work by Langrish *et al.* (1972), the authors found that the most important factor delaying successful innovation was the insufficient development of complementary technology. In the case of Korea, the Korean electronics firms are said have a high degree of dependence on the Japanese companies for sourcing their key components, a factor which is causing them to be one step behind the Japanese.[8]

Analysis of the above cases has highlighted a common trend: building technological capabilities in component generic technologies and key components allows firms to enter the next step in building competences at the systems level. This was the case for many firms. Thus the firms are pursuing a strategy of technological vertical integration, starting from component generic technologies and key components, and moving onto systems. In the case of Sumitomo Electric, the company began by building capabilities in III–V semiconductors, followed by optical fibres and have branched into systems, and key components. Having developed capabilities in key components, another factor which the firms considered was to become a merchant maker. Selling key components on the open market allows cost reduction and economies of scale. System divisions can choose the most favourable key component and can decide whether to use internally developed key components or purchase them on the open market. Such open competition allows the firm's own key components to become more competitive. However, some firms stressed that important key components might not be sold on the open market, in order not to give away their competitive edge. Some firms argued that they would have a natural lead time advantage anyway[9], so it would be safe to sell key components on the open market.

## 6.4  SHIFTING TRAJECTORIES: KEY COMPONENTS AND ECONOMIES OF SCOPE

As the two types of technological linkage maps show, economies of scope are achieved by the emergence of several sub-trajectories in parallel for a key component, arising because of changed development aims. For example, two types of AlGaAs semiconductor lasers emerged in several firms (e.g. NEC, Fujitsu, Siemens, Hitachi: Tables 6.9(b), 6.10(b), 6.12(b), 6.18(b)) one for communications and a high powered one suitable for optical disk and laser beam applications, which appeared later.[10] Although the first type of laser for communications was soon replaced by indium type semiconductor lasers around 1977, the technological accumulation was not lost. The firms were able to capitalize on what they had learned in the process of developing the first type of high powered lasers in developing the second type, based on the same material, which led to totally new applications.

Figure 6.2 shows the case of one company which began to work on semiconductor lasers in 1964. Several months after the discovery of room temperature oscillation of semiconductor lasers at Bell Labs in 1970, the company achieved the same. Initially, they had been working on gallium arsenide (GaAs) material in the 0.83 μm wavelength. Although the initial discovery was a major breakthrough, the semiconductor laser was only able to emit light for several seconds. It was not a simple task to make a semiconductor laser which lasted several thousand hours in order to pave the way for commercialization. During the next five years, improving life and finding out the causes of defects were the most pressing agenda. This was followed by other criteria such as obtaining single mode oscillation and improving quantum efficiency. In 1978, research effort was shifted to work on a different material based on indium gallium arsenide phosphide (InGaAsP) in the 1.3 μm wavelength region. Although the research using GaAs for communications was terminated, they were able to make use of the technological know-how gained by continuing research on CD applications based on the same material. The company exited early from the semiconductor laser market for CD application since the competition was fierce, but they decided to compete in the semiconductor laser market for optical data storage application for computers. Research on optical communication which began in 1967 reached the stage of conducting trials with NTT in 1977, and finally led to successful commercialization in the 1980s.

*Figure* 6.2 An example of shifting technological trajectories of semiconductor lasers

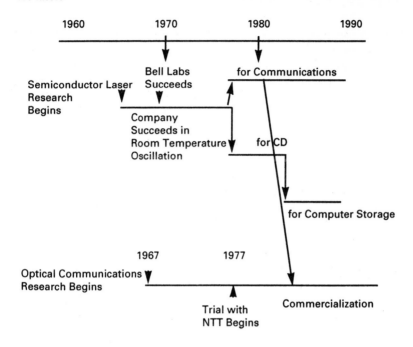

This example shows that competence building takes a long time, often spanning more than a decade or longer, and involves painstaking trial and error and organizational learning. This company was able to learn from their failures and were able to accumulate their technological bases successfully, and at the same time exploit technological synergies. The technological know-how and expertise enabled them to move from optical communications to CD players and other applications, such as optical data storage and laser beam printers, enabling the company to shift trajectories. This example also shows that while the initial technological driving force was in communications, subsequently the firm was able to use its capabilities in areas other than communications, i.e. in consumer and industrial electronics. Thus, while the initial phase of competence building centres in areas related to the firms' core businesses, later on, firms may be able to capitalize on the competences built to strengthen other business areas.

Similarly, in firms such as Sony, Sharp, Hitachi (Tables 6.15(b), 6.16(b), 6.18(b)) the technological know-how accumulated for developing semiconductor lasers for CD players or optical storage was used to develop other types of lasers such as high power lasers and visible lasers, thus opening up opportunities to enter new high value added applications. It was discussed earlier, in the case of Sharp and Sony, how optoelectronics supported the corporate goals of expanding the firms' businesses in non-consumer related areas. Such a technological strategy leads to increased earnings from development efforts.

In the field of optical communications, companies belonging to group 1 which were communications-related, were able to build on their accumulated competences in communications and IC technologies. Pulse code modulation was invented at STC in the 1930s. The company had been a leader in coaxial cable technology. They also had know-how in microwave communications. Having capabilities in these fields was a major advantage for building competences in optical communications. At STC, the earliest application of LEDs was for military applications, such as range finding and ship-to-ship communication. STC was able to build on this expertise to move into optical communications using its accumulated know-how in communications. Similarly, NEC had accumulated competences in digital communication systems, coaxial PCM, network control, transmission protocol, prior to entering the field of optical communications.

The technological linkage maps for LCDs also highlight economies of scope and incremental learning. LCDs can be classified into two types: the earlier version called simple matrix and the more sophisticated version which appeared more recently called active matrix type displays, whose individual pixels are controlled by TFTs (thin film transistors). Japanese firms have moved from simple to active matrix type displays; from low end applications such as watches and calculators, to word processors, first with only text, then with text and graphics and finally to TVs and projection displays, thus moving more upmarket as shown in Tables 6.16(c), 6.17(c), and 6.18(c). In contrast, GEC, for example, stayed in niche markets such as LCD for avionic displays (Table 6.13(b)) and did not effectively exploit economies of scope. Similarly, although STC had started to work on LCDs in the 1970s and commercialized them for special applications[11], they decided not to enter the mass market in 1985.

Kodama (1991) argues that the liquid crystal display is an example of technological fusion. Firms have been able to achieve economies of technology along these trajectories. The accumulated competences in the various underlying technologies form the foundation in conjunction with other emerging technologies to develop new higher value added systems.

This section has highlighted that competence building has been related to a large extent with technological accumulation and organizational learning. Also, maintaining an effective vertical linkage allows firms to diversify horizontally since firms may capitalize on economies of scope, and are able to make use of the accumulated technological bases to enter new markets via high value added systems.

## 6.5 CONCLUSION

This chapter demonstrated the existence of the linkages between component generic technologies, key components and systems. The statistical analysis showed that the coefficients of variance of component generic technologies and key components were much smaller than those related to products and systems. Through the use of technological linkage maps, the existence of the interlinkages was confirmed. The maps showed that TYPE I maps were similar for most firms independent of the group, while TYPE II maps were different according to the group. TYPE I maps indicate the generic component technologies which are used to develop key components, and which are common across a range of components. TYPE II maps which show the relation between key components and systems on the other hand, reflect the business interests of firms and hence vary between one another. However, TYPE II maps belonging to the same group showed similar features. This was because there were common business interests of the companies in each different group. It was thus possible to show empirically the concept of generic technologies which are the basic underlying technologies forming the foundation of a range of key components. These generic technologies are important for all firms irrespective of their final product markets.

This chapter also examined the nature of the search process and incremental learning. By examining the different types of papers related to semiconductor lasers over a 13 year period, it was found that in most firms, there was a decrease in experimental-type papers accompanied by a rise in papers marking 'new developments' or 'practical applications'. Such an existence of a wedge pattern for most firms confirmed the hypotheses that competence building is a long cumulative process characterised by trial and error and experimentation, which may eventually lead to fruitful outcomes. We noted a different wedge pattern between Japanese and European companies.

While competence building is strongly related to path dependency, it also provides opportunities for the firm to branch into new, more high value added areas. In other words, when competence building reaches a certain stage, it allows firms to shift their strategies. The case studies of the development of semiconductor lasers and liquid crystal displays showed that firms were gradually able to accumulate their technological bases and achieve economies of scope.

# 7 Conclusions

The central aim of this book has been to examine the dynamics of the competence-building process within firms. The principal argument has been that, in the development of technological competence, firms will view a radically new opportunity offered by a new technology such as optoelectronics from different angles, shaped to a large extent by their differing accumulated technological bases and business interests. Competence building is a painstaking and long process, entailing uncertainty and trial and error and it requires continuous learning. Several factors affect the rate and direction of competence building. Chief among them are the following: previous core business activities and links with main customers; top management strategy and the evolution of the R&D organization; government policy; management of the linkage between systems, key components and component generic technologies; organizational learning; and economies of scope.

We approached the problem using the framework of Nelson and Winter's (1982) evolutionary model. The concepts underlying the study were derived from the literature on the notion of competence. Competences embody intangible assets, skills and creative resources to generate new technological capabilities which are accumulated over time.[1] We assumed that the different types of competences, such as those related to technology, marketing and finance, could be considered independently. In this study, we focused on technological competence, which is a vitally important area of competence for firms in the hi-tech or science-based sector. Technological competence, then, involves the capacity to assimilate radically new technological oppotunities, deploy technological capabilities and to expand the range of technological capabilities. There are three levels to consider when discussing competence building. At the lowest level lie the component generic technologies, which are the basic enabling technologies for a company. These are combined with other generic technologies to produce key components (the second level) which are in turn used in conjunction with other generic technologies to develop a broad range of value added systems and products (the third level).

195

The book set out to examine the competence-building process in an international sample of 11 companies engaged in the field of optoelectronics, between 1977 and 1989.

## 7.1 MAIN FINDINGS

### Measurement of Competences

A novel technique to measure technological competence was developed, using mainly publicly available data other than performance-related data such as sales and market shares. In other words, we were able to measure the competence-building process dynamically, on an *ex ante* basis. Optoelectronics was divided into eight sub-areas. The three measures of assessing competences, using bibliometric INSPEC data, US patent data and interview data are generally consistent with one another, and together enable an important, intangible phenomenon to be measured. The various measures show that the areas of optoelectronics on which each firm has been concentrating (as measured by scientific publications and patents) are clearly related to one another. In order to capture the dynamics of competence building, IFTI (Intra-Firm Technology Index) figures based on publication and patent data were calculated over three periods. Time series data enabled areas in which firms were building competences to be highlighted. The use of these publication and patent data coupled with the firms' self-assessments enabled us to gain an understanding of the areas of relative strength and weakness in each company. The analysis showed that, within large firms, there are variations in the patterns of competence building depending on the companies' prior capabilities. The areas of strength and weakness within optoelectronics can be used to group the firms into three categories: (1) communications-driven (Fujitsu, NEC, Siemens, STC, GEC, Sumitomo Electric); (2) consumer and industrial electronics-driven (Toshiba, Sharp, Sony, Philips); and (3) evenly balanced firms (Hitachi). Independent of that analysis, the statistical technique of cluster analysis was used to categorize the firms using INSPEC and patent data. Similar results were obtained. Thus it was possible to provide empirical evidence on the emergence of firm specific trajectories.

The rate and direction of competence building is affected both by past historical and current forces which are more amenable to management action. Furthermore, competence building can be planned strategically with a long-term view.

**Previous Core Businesses and Links with Customers**

Competence building is strongly related to the firm's past accomplishments. The notions of path dependence and technological cumulativeness have a strong foundation. Competence building centres in key areas to enhance the firm's core capabilities. Firms in group 1 are those where optoelectronics-related competence building centred on optical communications. These firms had close ties with the national PTTs, although most also benefited from close ties with other public sector industries such as broadcasting, electric utilities, energy, transportation and defence. The PTTs played a significant role in acting as a technological driver for these firms. GEC's areas of strengths, for example, are closely related to its core business domains in defence systems and communication. Areas of weakness of these firms lie mostly in consumer electronics related areas. Firms in group 2 have a common feature in the way their optoelectronics-related competence building was driven primarily by consumer and industrial electronics rather than communications. These firms lacked the experience of building capabilities in communications systems. They have been able to build their optoelectronics competences in areas related to such as audio visual products, broadcasting equipment and IT. Hitachi is the only firm surveyed in group 3, and its optoelectronics-related competences are equally balanced between optical communications on the one hand, and consumer and industrial electronics on the other, reflecting its broad business activities.

**Strategy and R&D Organization**

Understanding and support by top management is essential for building competences. It is possible to locate the optoelectronics-related competence-building process of firms within the framework of the overall strategies formulated by top management. Optoelectronics offered a window of opportunity, allowing firms to pursue their long-term goals. For example, NEC has been pursuing a 'C&C' (Computers and Communications) strategy, while Toshiba has been driven by its triple 'I' strategy of 'Information, Intelligence and

Integration'. Sony and Sharp have been trying to strengthen their non-consumer electronics-related businesses, and their optoelectronics-related competence building could be seen in the light of this strategy. Top management of Japanese firms realized early on the importance of building technological capabilities in key components and component generic technologies. Indeed, the importance of core technologies was stressed in some firms as early as the 1950s and 1960s.

Analysis of top management vision and strategy confirms that the Japanese firms' strategies have been generally coherent, resulting in organizational changes and other actions taking place over a long time to improve the climate for building technological capabilities. The common feature of the two British firms investigated here is that they tended to have relatively short-term horizons, and both witnessed many changes in strategy. They were also rather risk averse. The example of LCDs shows that, although these two firms have built certain competences, they decided to stay in safe niche market segments (especially in defence) and top management decided not to invest in volume production for applications such as PC displays.

Top management strategy affects the evolution of the R&D organization, which in turn affects the competence-building process. This is consistent with Chandler's (1962) point of view that structure follows strategy. However, Marengo's (1991) argument that a firm's set of competences are shaped by organizational structure also seems to hold. The example of Hitachi shows that structure can affect competence building, but that strategy can in turn affect structure. The evolution of the organization of R&D has been very similar for the firms in group 2, which took many years gradually to form a three-layer R&D organization (with a central research laboratory, applications-oriented research laboratories and development units). During the 1980s, the R&D organizations became driven by core technologies. Integration has been taking place in Japanese firms, linking the activities of the central research laboratories conducting basic technology development, the applications laboratories defining broad application possibilities, and the development units developing specific products for market needs. While some difference exists among the firms in group 1, in general they show a similar pattern, gradually forming a core technology-driven system. In several firms, the R&D organization was further changed to integrate groups working on devices, component generic technologies and systems, thereby achieving vertical device integration.

**Role of Government Policy**

In Japan, NTT played a significant role as a technological driver, supporting the optoelectronics-related competence building of firms in group 1. In comparison, the MITI-led national optoelectronics project began late, just when the various components and systems were about to be commercialized, and did not make a significant contribution to firms' competence building. However, MITI's role can be seen from another perspective, in that it signalled to the world that optoelectronics had been chosen as a national project, spurring activities across "outsider" firms which had not been invited to join the project.

In contrast in the UK, the MoD and the British Post Office, like NTT, played an important role both as a technological driver, and in funding the research in the early stages of development.

**Convergence in Competences**

While the previous two sections have described the historical forces affecting the competence-building process, this section will describe current forces, looking at how the interlinkages between the three levels, organizational learning, and the organizational routines influence competence building. In firms facing similar technological opportunities, some factors lead to diversity in the directions of companies' technological search: learning in early stages, differing core businesses, differing management strategies. This is reflected in differences amongst firms in the priorities accorded to the various sub-fields within optoelectronics. However, in key components and their generic technologies, all firms allocate a substantial share of technical resources to their development. In Chapter 6 two approaches were used to analyze the relationship between systems, key components and component generic technologies. The first used statistical techniques. There was less variability amongst firms in the priority accorded to key components such as semiconductor lasers and their generic technologies than in other sub-fields of optoelectronics. Large values were measured for systems and products, such as optical disks, LCDs and communications systems. Thus the results provided empirical evidence for the concept of generic, pervasive technologies which are common areas in which all firms have been working, and which form the foundation of a variety of components used in different product and system applications. When we examined variation over time, then the variances across firms were larger for all technological fields in the earlier periods, compared to later periods, suggesting convergence of competences. The second approach was based on generating technological linkage maps from INSPEC and interview data for each firm. One type of map (TYPE I) shows the interlinkages between the component generic technologies and the key components. The second type of map (TYPE II) shows the interlinkages between key components, other generic technologies and systems, based on the three-level model above. A comparison of TYPE I maps across firms shows that they have many common features. These maps highlight the common areas in which firms have been building capabilities, irrespective of the final product markets. A comparison of TYPE II maps shows that there are certain inter-firm differences. However, common features can be seen among firms in the same group, reflecting the common business interests of the companies in each group. We can show through these interlinkage maps that the model, as described, has a strong empirical foundation. Analysis of TYPE I maps shows that some generic component technologies are more pervasive than others. It further indicates the long time span, often a decade or longer, during which certain companies have been working to develop these generic technologies.

In some firms, technological investments are made with a long-term view to building technological capabilities. Building capabilities in component generic technologies and key components, and managing the effective linkage between the three levels, offers firms strategic advantage over firms which buy in key components.

## Search Trajectories and Learning

In Chapter 6 the notion of search trajectories was tested using statistical techniques to measure the degree of diversification evident in the INSPEC and patent data. Firms search over a broad range of possibilities in basic and applied research and a rather narrower range in technology development. In the early phase of competence building, firms explore a broad range in order to deepen their knowledge of the field. Results of other statistical tests show that the search trajectory becomes more focused in later periods. As they accumulate knowledge, firms are able to narrow down their search path into areas which would strengthen their core businesses, through a painstaking learning process.

Learning is an activity which takes time and effort, involves trial and error and is far from costless. An analysis of the trajectory for one of the most important key components, the semiconductor laser, was made for each firm using INSPEC bibliometric data (Chapter 6). Over a 13-year period, we found that in most firms there was a decline in experimental-type papers, which was accompanied by a rise in papers related to new product development and practical applications. Such an existence of a wedge shaped pattern confirmed our proposotion that competence building is a long, cumulative process characterized by trial and error and experimentation, which may eventually lead to fruitful outcomes.

This study has shown that a firm cannot continue to build competences on their existing range of technologies alone. Examples of the development of optical fibres, and CD players confirmed that at some point in time, firms have to acquire certain new technologies and techniques from scratch. In the case of optical fibres, learning the properties of glass was totally new for Sumitomo Electric. However, they were able to use some existing techniques such as cable coating. Competence building entails a combination of cumulative technological advance as well as acquisition of new skills.

While competence building is constrained by path dependence, it also provides opportunities for firms to branch into new areas, allowing firms to shift their trajectories. The technological linkage maps demonstrate the emergence of several sub-trajectories for a key component. Economies of scope are achieved by the emergence of several sub-trajectories in parallel for a key component, arising because of changed development aims. In several firms, although the first type of semiconductor laser intended for use in communications had to be abandoned after seven or eight years, the technological accumulation was not completely lost. The firms were able to capitalize on economies of scope, using the know-how gained to develop semiconductor lasers for new applications, such as optical storage for computers and laser beam printers. These aspects are linked to our argument that vertical linkage can lead to horizontal diversification, which further underlines the importance of mastering key component technologies.

**Competence-building Patterns**

The competence-building pattern, during the initial, mid-, and later phases has been examined. In most cases, the research was begun spontaneously by researchers who thought it was an exciting new field which might become important for the company in the future. At that stage, they had no clear applications in mind. For example, at Sumitomo Electric, when research on gallium arsenide started, researchers wanted to discover what kind of a device could be made with it using the property of being much faster than silicon. The material, which was later used in key components for optical communications, was not at first considered for that application. It was only eight or nine years later that applications in optical communication emerged. Hence, the curiosity of company researchers plays an important role in the initial phase of competence building. A feedback mechanism exists which rewards the initial search process by researchers with top management support. Thus, the progression from discovery in the laboratories will often lead to top management awareness and a recognition of the need to support the research, provided it falls within the domain set by top management's long-term strategy. In other cases, the initial entry into optoelectronics was due to internal demand; or it was a necessary step to maintain links with main customers. In short, both technology-push and market-pull are influential factors in the competence building process.

During the competence-building phase, some firms have been able to take advantage of opportunities which have come about by chance. Strategic planning, nevertheless plays an important role in supporting long-term research and in setting up an organizational structure conducive to competence building.

**Competence-sustaining Routines**

Once competences have been created, it is all too easy for firms to lose them unless these competences are continuously upgraded. Some firms have developed explicit methods to sustain competences. In such organizations, there are often effective horizontal linkages across functional units, and vertical linkages across systems, key components and the component generic technology units. Most Japanese firms have a system similar to Tokken (a name invented at Hitachi, meaning Special Project) allowing the mobilization of resources across the organization to develop specific core technologies and products. Another way to improve the horizontal linkages is to have 'core technology champions'. The champion is in charge of monitoring world-wide trends, attending conferences, coordinating various optoelectronics activities in the firm, clarifying their respective responsibilities.

## 7.2 IMPLICATIONS AND SUGGESTIONS FOR FUTURE RESEARCH

One critically important finding from this study is that in the area of optoelectronics, in-house development has been the primary mode of competence building, especially in the early and middle phases of development. Firms have been building capabilities in generic technologies for one to two decades, incrementally adding to their technological bases, through trial and error and organizational learning. After the initial discovery of semiconductor lasers in the laboratory, it took 10 years to reach the stage of commercialization. It is a debatable point as to why firms choose to develop their competences by themselves. Williamson (1975) claims that integration may permit transaction economies. In other words, firms may be able to save on transaction costs by internalising the competence-building process. Furthermore, he points out that the supplier may respond opportunistically. The research reported here has shown that failure to build competences in component generic technologies and instead relying on buying in components from others, may run the risk that the company fails to be competitive in systems and products in the long run. As shown in Chapter 6, mastery of component generic technologies led to fruitful outcomes in several unforeseen areas. Thus, maintaining effective vertical technological linkages allows firms to capitalize on economies of scope, and enables them to diversify horizontally. In other words, one of the important findings of this study is that building capabilities in the upstream end of component generic technologies and key components, through long term in-house effort, gives rise to organizational learning, which can then be applied in other new areas at the downstream end of the market.

The immediate question following from this research is whether optoelectronics is a special case, or whether the findings are generalizable. It would be logical to carry out research examining whether this competence-building model and the technique to assess competences can be applied to other fields, such as precision instruments or office equipment.[2]

This work has focused on the aspect of technological competence building. We have not been able to extend our argument to link competence building with competitiveness. To undertake research into the relationship between technological competence and the firm's competitiveness is practical, and potentially useful. For example, one can examine the linkage between technological

competence and its product innovation competence. Strong technological competence may be a necessary condition for a firm to be
competitive, but not a sufficient one.

We examined how the scope of competences built changed over
time. The statistical tests using Herfindahl indices revealed that in
some firms there was divergence of competences within optoelectronics, while in others there was convergence. An interesting topic of investigation is to explore the variation of scope of competence
building in relation to the evolution of firms.

Given the pace of change in the 'high-tech' industries, firms have
to wrestle with the increasing complexity of products and technologies. Firms may not be able to build competences by themselves and
may have to rely more on external linkages in the form of joint ventures or technological collaboration. In recent years, new forms of
competence building, based on external linkages have begun to take
place, as we saw in Chapter 4. Innovation is becoming more of a
multi-firm networking process (Rothwell, 1992). Firms may have to
manage effective vertical linkages using key components or generic
technologies developed by others or working with other organizations. Research into these matters might provide insights into the
dynamics of competence building, during the 1990s.

An issue which needs further investigation and one on which this
research has shed some light is the innovation process. Langrish *et
al.* (1972) had earlier concluded that simple linear models of innovation were unrealistic, noting that the sources of innovation were
multiple. Nelson and Winter (1977) proposed a 'selection environment' model. In 1979, Mowery and Rosenberg provided a critical
review of some innovation models. They pointed out that the
concept of demand used in many of the studies is quite vague.
Rothwell (1991) summarized the five models of industrial innovation, starting from the simple linear sequential technology-push and
the need-pull models of the 1960s and early 1970s and moving to the
interactive coupling model which was dominant until the early
1980s. The fourth generation model is the one which considers innovation as a parallel process involving R&D, prototyping, manufacturing and so on. The fifth generation model is based on a close
strategic integration between collaborating companies. This work
has shown that the emerging trajectories were affected by both technology-push and demand-pull-related elements. Furthermore, competence building through in-house effort was very important. The
selection environment model, therefore, has to be reassessed. This

research has also shown the existence of a strong internal network linking R&D and business units. The systemic innovation model based on linkages between generic technologies, key components and systems outlined in Chapter 2 can be modified to add the dimension of economies of scope, which was seen through the emergence of several sub-trajectories for the same type of key component leading to new applications. A new innovation model could be constructed on the basis of combining the selection environment model and the modified systems innovation model.

This research revealed cases where, although some companies were early starters, failed to become market leaders and eventually lost their competence, while other firms were late starters but managed to catch up in key areas. Organizational routines to build and sustain competences are crucially important. Although several forms of such routines were identified, it was beyond the scope of this book to go into greater depth. Research into these matters, possibly through in-depth case studies, is needed especially for making policy recommendations to firms.

We have seen how industrial R&D laboratories are the means whereby large companies can learn and assimilate major technological discontinuities, as Pavitt and Patel (1991) have pointed out. The importance of learning in R&D has been highlighted here. Some empirical evidence on the coupling of search paths with organizational learning was produced. At the same time, certain trends which may be a characteristic of the Japanese innovation system were noted, in particular that the search path was consistently more focused than in the West (Chapter 6). Only detailed investigation can establish the source of such national specificities in the innovation process.

Technology strategy has often been discussed in the past in terms of simple choices such as whether to be an innovator or a follower, a low cost producer or the adopter of a niche strategy. The evidence of this study points to the need to view technology strategy from another perspective, centred on the dynamics of competence building.

# Appendix A Systemic Concept used at Hitachi

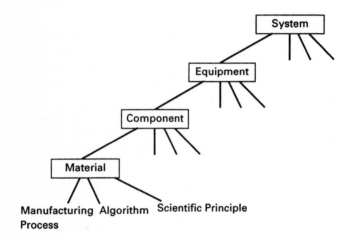

*Source*: Hitachi Research Institute (April 1983).

# Appendix B.1
# Optoelectronics-related Production in Japan (million yen)

| Optical components | 1987 | 1988 | 1989 |
|---|---|---|---|
| Semiconductor Laser | 34 528 | 29 411 | 33 569 |
| Gas Laser | 7 282 | 8 512 | 11 024 |
| Solid State Laser | 2 229 | 2 976 | 3 316 |
| LED (for Comm) | 6 232 | 4 391 | 4 716 |
| LED (for Display) | 59 968 | 64 611 | 76 605 |
| Other Lasers | 63 | 15 | 18 |
| Light-Receiving | 66 963 | 89 600 | 97 247 |
| Compound Elements | 36 714 | 46 294 | 47 950 |
| Display Component | 97 734 | 112 298 | 161 673 |
| Solar Battery | 10 352 | 9 733 | 11 486 |
| Optical Fibre | 71 147 | 54 671 | 65 724 |
| Image Fibres | 5 766 | 6 889 | 9 337 |
| Optical Connectors | 10 976 | 20 658 | 18 507 |
| Optical Active Elements | 6 545 | 6 828 | 7 168 |
| Other Components | 42 521 | 15 455 | 18 133 |
| Sub-total | 459 020 | 472 342 | 566 473 |

| Products | 1987 | 1988 | 1989 |
|---|---|---|---|
| Optical Transmission Equipment | 56 968 | 60 729 | 65 168 |
| Optical Measuring | 17 561 | 18 776 | 20 864 |
| Construction | 9 102 | 5 924 | 7 582 |
| Optical Sensor | 49 061 | 48 977 | 51 728 |
| Optical Disks and CD Equipment | 360 083 | 454 418 | 527 698 |
| Video Disk Equipment | 58 827 | 89 779 | 130 019 |
| CD-ROM Equipment | 34 046 | 41 596 | 62 458 |
| Rewritable Optical Disk | 470 | 2 010 | 9 075 |
| CD | 152 942 | 220 921 | 309 349 |
| Video Disk | 87 180 | 100 366 | 112 409 |

| CD-ROM | 2 553 | 4 163 | 8 503 |
|---|---|---|---|
| Input Output Equipment | | | |
| Printers, Copiers | 180 834 | 301 908 | 331 246 |
| Bar Code Readers | 7 971 | 8 597 | 10 406 |
| Others | 11 829 | 15 901 | 18 608 |
| Display Equipment | 1 832 | 2 273 | 3 795 |
| Medical Laser | 5 328 | 5 026 | 5 277 |
| Laser Applied Equipment | 37 161 | 52 322 | 66 008 |
| Printing Equipment | 17 818 | 22 980 | 22 949 |
| Other Equipment | 414 | 401 | 468 |
| Sub-total | 1091 978 | 1457 067 | 1763 610 |

| *Optical communication system* | *1987* | *1988* | *1989* |
|---|---|---|---|
| Public Communication | 71 652 | 87 956 | 96 070 |
| Private User | 66 206 | 64 602 | 67 604 |
| Other Systems | 1 760 | 4 037 | 4 503 |
| Sub-total | 139 618 | 156 595 | 168 177 |

| Total | 1690 616 | 2086 004 | 2498 260 |
|---|---|---|---|

*Source*: OITDA (Optoelectronic Industry and Technology Development Association) *Optonews*, 1990.

# Appendix B.2
# Optoelectronics-related Production in Japan (as % of Value)

| Optical components | 1987 | 1988 | 1989 |
|---|---|---|---|
| Semiconductor Laser | 2.04 | 1.41 | 1.34 |
| Gas Laser | 0.43 | 0.41 | 0.44 |
| Solid State Laser | 0.13 | 0.14 | 0.13 |
| LED (for Comm) | 0.37 | 0.21 | 0.19 |
| LED (for Display) | 3.55 | 3.10 | 3.07 |
| Other Lasers | 0.0 | 0.0 | 0.0 |
| Light-Receiving | 3.96 | 4.30 | 3.89 |
| Compound Elements | 2.17 | 2.22 | 1.92 |
| Display Component | 5.78 | 5.38 | 6.47 |
| Solar Battery | 0.61 | 0.47 | 0.46 |
| Optical Fibre | 4.21 | 2.62 | 2.63 |
| Image Fibres | 0.34 | 0.33 | 0.37 |
| Optical Connectors | 0.65 | 0.99 | 0.74 |
| Optical Active Element | 0.39 | 0.33 | 0.29 |
| Other Components | 2.52 | 0.74 | 0.73 |
| Sub-total | 27.15 | 22.64 | 22.67 |

| Products | 1987 | 1988 | 1989 |
|---|---|---|---|
| Optical Transmission Equipment | 3.37 | 2.91 | 2.61 |
| Optical Measuring | 1.04 | 0.90 | 0.84 |
| Construction | 0.54 | 0.28 | 0.30 |
| Optical Sensor | 2.90 | 2.35 | 2.07 |
| Optical Disks and CD Equipment | 21.30 | 21.78 | 21.12 |
| Video Disk Equipment | 3.48 | 4.30 | 5.20 |
| CD-ROM Equipment | 2.01 | 1.99 | 2.50 |
| Rewritable Optical Disk | 0.03 | 0.10 | 0.36 |
| CD | 9.05 | 10.59 | 12.38 |
| Video Disk | 5.16 | 4.81 | 4.50 |
| CD-ROM | 0.15 | 0.20 | 0.34 |

Input Output Equipment

| Printers, Copiers | 10.70 | 14.47 | 13.26 |
|---|---|---|---|
| Bar Code Readers | 0.47 | 0.41 | 0.42 |
| Others | 0.70 | 0.76 | 0.74 |
| Display Equipment | 0.11 | 0.11 | 0.15 |
| Medical Laser | 0.32 | 0.24 | 0.21 |
| Laser Applied Equipment | 2.20 | 2.51 | 2.64 |
| Printing Equipment | 1.05 | 1.10 | 0.92 |
| Other Equipment | 0.02 | 0.02 | 0.02 |
| Sub-total | 64.59 | 69.85 | 70.59 |

| *Optical communication system* | 1987 | 1988 | 1989 |
|---|---|---|---|
| Public Communication | 4.24 | 4.22 | 3.85 |
| Private User | 3.92 | 3.10 | 2.71 |
| Other Systems | 0.10 | 0.19 | 0.18 |
| Sub-total | 8.26 | 7.51 | 6.73 |
| Total | 100.00 | 100.00 | 100.00 |

*Source*: OITDA (Optoelectronics Industry and Technology Development Association), *Optonews* 1990.

# Appendix C    Trend of Liquid Crystal Displays

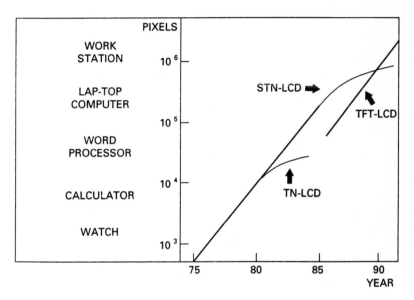

*Source*: Hitachi Research Laboratory (December 1990).
*Notes*:  TN-LCD = Twisted Nematic LCD
  STN-LCD = Super Twisted Nematic LCD
  TFT-LCD = Thin Film Transistor LCD

# Appendix D   Breakdown of Sales by Sectors in Firms (as % of values)

| Toshiba | 1965 | 1975 | 1985 |
|---|---|---|---|
| Consumer Electronics | 51 | 41 | 33 |
| Heavy Industrial | 33 | 37 | 32 |
| IT, Communications | 16 | 22 | 35 |
| Sales (100 m yen) | 2 213 | 8 939 | 25 196 |
| R&D as % of Sales | 1 | 1.2 | 6.6 |

| Hitachi | 1970 | 1975 | 1980 | 1985 |
|---|---|---|---|---|
| Heavy Industrial | 24.0 | 23.4 | 24.8 | 20.7 |
| Consumer Electronics | 29.3 | 27.4 | 23.7 | 13.8 |
| Info, Comm. Systems and Devices | 19.9 | 20.6 | 27.0 | 36.0 |
| Industrial Machinery | 16.8 | 15.5 | 11.4 | 9.3 |
| Traffic Equipment | 10.0 | 13.2 | 13.1 | 10.3 |
| Sales (100 m Yen) | 8 072 | 11 106 | 19 425 | 30 034 |
| R&D as % of Sales | 5.1 | 5.5 | 5.9 | 8.1 |

| NEC | 1970 | 1975 | 1980 |
|---|---|---|---|
| Communication Eqpt | 57.6 | 51.2 | 39.0 |
| Devices | 20.3 | 16.4 | 24.5 |
| Electronics | 20.3 | 23.4 | 26.8 |
| Home Electr. and others | 6.0 | 9.0 | 9.5 |
| Sales (100 m yen) | 1 177 | 4 275 | 8 929 |
| R&D as % of Sales | N.A. | 4.3 | 5.5 |
| Private Sector (%) | | 39.6 | 47.5 |
| Public Sector | | 33.2 | 22.2 |
| Export | | 27.2 | 30.4 |

*Note*: Home Electr. is equivalent to consumer electronics.
Electronics includes computers.

| *Sumitomo Electric* | 1978 | 1980 | 1985 |
|---|---|---|---|
| Cables | 62 | 58 | 55 |
| Special Wire Alloys | 17 | 16 | 18 |
| Brakes | 5 | 5 | 5 |
| Equipment | 4 | 3 | 3 |
| Construction Work | 6 | 12 | 5 |
| Hybrid Products | 3 | 2 | 5 |
| New Businesses | 3 | 4 | 9 |
| Sales (100 m yen) | 3 110 | 4 289 | 5 321 |
| R&D as % of Sales | 2.1 | 2.0 | 3.4 |
| Export (%) | 13 | 19 | 13 |

| *Fujitsu* | 1965 | 1975 | 1980 | 1985 |
|---|---|---|---|---|
| Communications | 59.0 | 25.5 | 20.9 | 15.7 |
| IT, Computers | 27.4 | 71.9 | 65.8 | 71.9 |
| Devices | 13.6 | 2.6 | 13.3 | 12.4 |
| Sales (100 m yen) | 438 | 2 883 | 5 817 | 14 295 |

| *Sharp* | 1965 | 1975 | 1980 | 1985 |
|---|---|---|---|---|
| TV/Audio | 52 | 40 | 45 | 45 |
| Home Electronics | 43 | 33 | 21 | 22 |
| IE and Devices | 6 | 27 | 34 | 33 |
| Sales (100 m yen) | 293 | 2 018 | 5 014 | 9 553 |
| R&D as % of Sales | N.A. | N.A. | 3.6 | 6.2 |

| *Sony* | 1981 | 1984 | 1987 |
|---|---|---|---|
| Video Equipment | 34.5 | 40.6 | 29.0 |
| Audio Equipment | 28.7 | 21.5 | 30.8 |
| TV | 26.1 | 23.6 | 20.3 |
| Other | 10.7 | 14.3 | 17.4 |
| Music, Film | | | 2.5 |
| Sales (100 m yen) | 10 510 | 12 615 | 15 552 |
| R&D as % of Sales | 5.9 | 7.9 | 8.3 |
| Overseas Sales (%) | 70.9 | 72.6 | 65.4 |
| Domestic Sales | 29.1 | 27.4 | 34.6 |

IE = Industrial Electronics; N.A. = Not Available.

| Siemens | 1979 | 1981 | 1983 |
|---|---|---|---|
| Components | 5 | 4.6 | 4.3 |
| Data Systems | 5 | 5.5 | 6.1 |
| Power Engineering and Automation | 23 | 26.6 | 28.1 |
| Electrical Installation | | 9.5 | 7.6 |
| Communications | 29 | 30.0 | 28.6 |
| Medical Engineering | 7 | 8.7 | 9.1 |
| Power Plant | 9 | 9 | 10.9 |
| OSRAM | | 3.8 | 3.5 |
| Other | | 2.3 | 1.8 |
| Sales (DM bn) | 28.0 | 34.6 | 39.5 |
| R&D as % of Sales | 9.6 | 9.5 | 9.0 |
| Export (%) | | 56 | 56 |

| GEC | 1979 | 1985 |
|---|---|---|
| Electronic Telecom Systems and Components | 37.7 | 31.6 |
| Telecom and Business Systems | – | 12.7 |
| Automation and Control | – | 7.9 |
| Medical Equipment | | 8.0 |
| Power Generation | 17.5 | 11.1 |
| Electrical Equipment | 14.3 | 13.8 |
| Consumer Electronics | 12.2 | 5.1 |
| Distribution | | 3.9 |
| Others | 4.0 | 6.0 |
| Sales (million £) | 2 941 | 5 892 |
| R&D as % of Sales | N.A. | 10.9 |
| Export (%) | | 22.1 |

*Telecom = Telecommunications*

| STC | 1979 | 1985 |
|---|---|---|
| Computers | | 54.6 |
| Telecommunications | 60 | 13.8 |
| International Communications | | 11.9 |
| Defence | | 4.8 |
| Components, Distribution | 40 | 14.8 |
| Sales (million £) | 437 | 1 896 |
| R&D as % of Sales | 4.1 | 8.1 |

*Note*: The figure for GEC Elec Systems and Compo in 1979 includes telecommunications and automation systems.
The figure for STC Telecom in 1979 includes International Communications and Electronics.

| Philips | 1985 |
|---|---|
| Lighting | 12.4 |
| Consumer Electronics | 26.3 |
| Domestic Appliances | 10.3 |
| Professional Products | 27.8 |
| Components | 18.1 |
| Miscellaneous | 5.1 |
| Sales (million F) | 60 045 |
| R&D as % of Sales | 6.7 |

# Appendix E.1 Summary of NEC

| Critical events | Organizational features | Role of other agents |
|---|---|---|
| **1899** – Founded as a joint venture of a Japanese firm and Western Electric (ATT) to import communications equipment | | **1899–1940** Became the main supplier of broadcasting and communications equipment to the Japanese government[1] |
| **1923** – Begins improving the imported comm. equipment | | **1923** – Automatic telephone exchanges made |
| **1930s** – Begins research on TVs ; ATT relinquishes control of Western Electric, resulting in NEC to become part of Sumitomo group | | |
| **1935** – Begins research on switching circuits | | |
| **1940s** – Begins research on semiconductors | **1940** – Research organization was similar to ATT's. R&D was separate from production | |
| | **1940s** – The organization was organized into functional units of production, sales, R&D etc. | |
| | | **1945** – Becomes a supplier to NTT of communications equipment, telephone exchanges, microwave circuits and systems |

216

| Critical events | Organizational features | Role of other agents |
|---|---|---|
| **1950s** – Computers developed for internal use | | |
| | **1951** – Closes R&D laboratory | **1951-** Electromagnetic waves liberalization enacted |
| **1953** – Spins off NEC Systems Construction, which constructs earth stations for satellite comm. | **1953** – R&D laboratory reopens | **1953** – Microwave communications equipment supplied to Tohoku Electric Power |
| **1957** – Begins research on digital comm. such as PCM | **1957** – Four main business units exist; Marketing and Sales transferred to the business units | |
| **1960** – Begins research on ICs[2] **1960s** – Top management realizes that NEC cannot rely on government contracts indefinitely. Aims to shift NEC towards serving the private sector[3] | **1960s** – Spins off production facilities | **1960** – Computers sold to Universities, government and other firms **1960s** – Sells microwave comm. equipment to Tokyo Electric power and other power companies **1960s** – Has links with Honeywell for developing computers |
| **1961** – Research on gas lasers (ruby lasers) begins | **1961** – R&D consists of basic research, Comm. Technology lab, Electrical equipment lab, Atomic Physics lab, Production Tech lab[4] | |
| | | **1962** – Hughes aircraft requests joint development of satellite communications technology[5] |

| Critical events | Organizational features | Role of other agents |
|---|---|---|

**1963** – Involved in the first TV space relay experiment between US and Japan

**1964** – Begins research on semiconductor lasers

**1964** – Kobayashi becomes president

**1965** – Zero Defect action carried out throughout the company

**1965** – Designates 2 main areas as NEC's core business domains, communications and electronics (computers)

**1965** – Withdraws from nuclear energy[6]

**1965** – Adopts SBU structure, having 14 business units, 3 development divisions

**1965** – Research which was previously split into basic research, communications and electric machinery research was merged

**1966** – President announces knowledge industry as NEC's corporate vision[7]

**1966** – R&D Evaluation system started

**1966** – Central Research Laboratory consists of Materials, Solids, Quantum Electronics, Communications, Infomation Processing, Prototyping divisions. In addition Production Engineering research lab exists[8]

**1967** – Begins research on optical communications

**1967**–Established Integrated Circuit as a business unit; Uenohara who had spent ten years at Bell Labs moves to NEC as director of the Central Research Laboratory

| *Critical events* | *Organizational features* | *Role of other agents* |
|---|---|---|
| | **1968** – New type of workers, called SE (Systems Engineers) appears, taking over the job previously done by non technical sales force | |
| | | **1969** – Joint development of Pulse code transmission carrier equipment (PCM) with NTT<br>**1969** – Develops optical fibres (self focusing glass fibre) jointly with Nippon Sheet Glass |
| **1970** – Electronic switching system completed<br>**1970** – Succeeds in room temperature oscillation of semiconductor lasers in the same year as Bell Labs<br>**early 1970s** – Begins research on LCDs but on a small scale | **1970s** – Builds up an effective dealer network for semiconductor chips and other products; Dealers grow from 190 in 1975 to 231 in 1980[9]<br>**1970s** – A keen promoter of optical comm. called Uchida who used to work on microwave comm.<br>**1971** – Electronic Device planning section set up<br>**1972** – R&D planning set up; Uenohara sets up a system of 'commisioned research' to establish close links between R&D and the business units | **1970** – Room Temperature oscillation of semiconductors at Bell Labs |

| Critical events | Organizational features | Role of other agents |
| --- | --- | --- |
| **1973** – Gets an award on analysis of defects of semiconductor lasers | | |
| **1974** – Launches ACOS computer | **1974** – Apart from the Central Research laboratory, Production Automation unit, Communication Network Promotion, Pollution Research, Automobile Electronics, Lasers development and Computing Centre exist | **1974–5** Joint research with Tokyo Electric Power, Kansai Electric Power on optical comm.; NTT decides to go ahead with optical communications |
| **1975** – Begins production of ICs in Ireland | **1975** – New R&D Centre completed | **1975** – Supplies optical comm. module for NTT |
| **1975–6** Importance of optical comm. firmly established | | **1975** – Collaborates with NTT on trial optical transmission. Succeeds in 400 M bit/s transmission over a distance of 5 km |
| **1976** – Distributed Processing architecture 'DINA' announced | | **1976–80** Participates in the national VLSI project |
| **1976–7** Terminates research on optical disks for 4–5 years | | |
| **1977** – C&C concept announced by the chairman Kobayashi[10] | | |
| **1977** – External sales ratio of devices including semiconductors reaches 84%[11] | | |
| **1977–8** Research effort on semiconductor lasers shifts to indium based material | | |

| Critical events | Organizational features | Role of other agents |
|---|---|---|
| **1978** – Acquires Electronics Arrays, a Silicon Valley IC manufacturer | **1978** – C&C advertising campaign to swtich corporate identity from being dependent on the public sector to that driven by private sector demand[12] | **1978** – Exports optical comm. system to Vista of Florida in US |
| | | **1979–85** Takes part in the National Optoelectronics Project. Works on short wavelength semiconductor lasers |
| **1980** – Organizes a (Computer and Communications) fair, featuring optoelectronics **1980s**–5 Strategic technical domains identified; material device, functional device, comm. systems, knowledge information, software | **1980** – R&D reorganized into R&D group consisting of C&C Research laboratory, Fundamental Research laboratory, Microelectronics Research lab, Optoelectronics Research lab, Software Engineering lab, Environmental lab, a Computing centre and a Production Engineering lab; R&D Planning section was established; Altogether employs 1500 scientists in 8 laboratories Within the Optoelectronics lab, device, system and material teams work together; The reorganization done to reflect NEC's C&C strategy | **1980** – Supplies a system for Argentine Intel |

| Critical events | Organizational features | Role of other agents |
|---|---|---|
| | Prior to this, optoelectronics research was done in the quantum device lab. | |
| | **1980s** – 3–4 Tokken type projects in progress | |
| | **1980s** – About 30% of R&D at the Central Research laboratory funded by the business unit. | |
| | R&D planning becomes an integral part of strategic planning. | |
| | **1982** – Fundamental technology lab split into basic research and microelectronics lab | **1983** – Wins a contract from Singapore Telecom and Thai Telecom to supply optical comm. system[13] |
| | **1983–5** Core technology program launched | |
| **1985** – Rapid catch up in development of LCDs through Tokken type project | | **1985** – Wins a contract from New Zealand PTT |
| **1987** – Develops short wavelength emitting laser suitable for use in optical disks | | |
| **1990** – Develops special VLSI for use in broad band exchange for ISDN | | **1990** – The breakdown of users in optical comm. shows that 70–80% is for public comm. (10% for LAN) |
| | | **1992** – MITI plans a national project on optical computing. |

[1]Kobayashi (1985).
[2]Kojima and Ikehata (1984).
[3]Uenohara (1987).
[4]Uenohara (1987).
[5]Ohmae (1984).
[6]Ohmae (1984).
[7]Kobayashi (1985).
[8]Kojima and Ikehata (1984).
[9]Ohmae (1984)
[10]Kobayashi (1985).
[11]Uenohara (1987).
[12]Ohmae (1984)
[13]*Hikari Sangyo Maker Souran* (Optical Industries Makers Directory) (1987).

# Appendix E.2 Summary of Fujitsu

| Critical events | Organizational features | Role of other agents |
| --- | --- | --- |
| **1935** – Fujitsu was founded as a spin off of the communications equipment division of Fuji Electric | | |
| | | **1950s** – NTT plans to set up a nation wide comm. infrastructure<br>**1950s** – Microwave technology transferred from Siemens[1] |
| **1953** – Begins to manufacture wireless comm. equipment | | |
| | | **1956** – Cross bar telephone exchange commercialized for NTT, becomes a cash cow<br>**1958** – Joins joint research on telephone exchanges with NTT and Hitachi which began in 1956 |
| **1959** – Fifth president Okada stresses the importance of core technologies | | |
| **1960s** – Prior to entry into optoelectronics. research on component generic technologies such as crystal growth carried out in the semiconductor materials research laboratory | **1960** – Doubles the intake of new recruits to 1200 | **Until mid 1960s** – Derives 70% of turnover from NTT |

| *Critical events* | *Organizational features* | *Role of other agents* |
|---|---|---|

**1961** – Entry into computers;
Adopts the strategy of developing its own technology rather than licensing foreign technology
**1962** – Begins research on semiconductors; Designates computers and electronics as two core business domains

**1961** – Two business units, Communications and Electronics (Computers) formed. Adopts SBU approach

**1962** – Fujitsu R&D Centre established through a merger of R&D activities undertaken in separate divisions.[2]
**1964** – Fujitsu is half the size of NEC, one tenth of IBM
**1968** – R&D Centre becomes an independent subsidiary to conduct research independently of short term business goals

**1968** – Acquires Kobe Industries which has expertise in semiconductors[3]

**1969** – First domestic computer made and sold to Universities and Banks
**1970** – The percentage of sales derived from IT and computers reaches 57% compared to 27% in 1965
**early 1970s** – Begins basic research on optoelectronics

**1971**–Collaborates with Hitachi on computers
**1972** – Begins to collaborate with Amdahl of the US. Adopts IBM compatible strategy; Learns from Amdahl

**1972** – Spins off numerical control (robot) division as an independent subsidiary called Fanuc

| *Critical events* | *Organizational features* | *Role of other agents* |
| --- | --- | --- |
| | | not only technologies, but aspects of R&D management |
| **1973** – Begins research on AlGaAs semiconductor lasers and LED for comm. | | **1973** – Concludes agreement with Furukawa Electric to re-license Corning's optical fibre technology |
| | | **1975** – Begins OEM supply of large mainframe computers to Amdahl |
| **1976** – Shifts to work on InGaAsP lasers. However know-how gained to develop AlGaAs lasers used for printers[4] | | |
| **1976–8** Begins research on Germanium and Indium based photodiodes | | |
| **1977** – Becomes a merchant maker of components; Until then they were for internal use and IC division was part of the computer group | | |
| **1979** – Overtakes IBM Japan in domestic sales of computers | **1979** – Sets up Optical Semiconductor devices laboratory in the Electron Device division to carry out research on optoelectronics | **1979–86** Takes part in the National Optoelectronics Project. Does work related to OEIC |
| **1979** – SAS structure laser developed but not commercialized since its performance is not good enough | | **1979–84** Collaborates with NTT to develop InGaAsP lasers and Ge photodiodes |
| **1980** – Achieves a world record in HEMT devices at room temperature, based on MBE | **1980** – SBU system adopted; the company consists of 3 business units of telecomm. computers and devices. Each unit has | **1980s** – InGaAsP lasers used in NTT's 400 M b/s trunk route |

| *Critical events* | *Organizational features* | *Role of other agents* |
|---|---|---|

technology.
Performance is 200
times better than
silicon devices

an admin and
Technology
Promotion division
which coordinates
technologies

**1981** – Experience
gained to develop
SAS lasers leads to
succesful development
and commercialization
of VSB lasers
**1981** – Ge photodiodes
transferred to
production

**1981** – Although ATT
had chosen Fujitsu to
install optical network
between Boston and
NY, they were
blocked by the US
government[7]

**1982** – World's first to
use VSB lasers in
submarine optical
comm.
**1983** – Technical
alliance with ICL
formed

**1983** – R&D
laboratory in Atsugi is
established. R&D
Centre consists of
Electron devices,
Electronic systems,
Materials lab in
Atsugi;
Communications and
Data Processing
systems laboratory in
Kawasaki. In addition
there is a Social
Information Science
laboratory.[6]
Altogether 1500
people engaged in
R&D; About a third
of the funding is
provided by the
divisions and the rest
by the head quarter
**1984** – Reorganized
the optoelectronics
activities in the
Materials laboratory,
leaving it to undertake
the basic research and

**1984** – External sales
ratio of devices
reaches 70%,
compared to 30% in
1977[5]

| Critical events | Organizational features | Role of other agents |
| --- | --- | --- |
| **1984** – Indium based photodiodes transferred to production **1986–7** – Commercializes DFB lasers **1988** – Company slogan 'Fresh Fujitsu' **1989** – Company slogan 'Exciting Fujitsu' | moving the respective epitaxial teams into the devices teams **1989** – LCD unit became attached to the headquarter **1990** – Fujitsu's sales is 2900 bn yen, compared to NEC's 3600 bn. Fujitsu is now one quarter of the size of IBM **1991** – Optical Interconnect promotion division created | **1989** – Spends about 200 bn yen acquiring ICL **1992** – MITI plans a national project on optical computing |

[1]Kojima and Ikehata (1984).
[2]Brochure on Fujitsu R&D, by Fujitsu.
[3]Kojima and Ikehata (1984).
[4]INSPEC Data.
[5]Kojima and Ikehata (1984).
[6]*Annual Report* (1989).
[7]Kojima and Ikehata (1984).

# Appendix E.3   Summary of Sumitomo Electric

| Critical events | Organizational features | Role of other agents |
|---|---|---|
| **1897** – Founded in Osaka as a member of the Sumitomo conglomerate (zaibatsu) whose origins date back to the 18th century, based on copper mining. A company to make copper cables and wires, Sumitomo Electric inherits the characterisitics of Sumitomo management of not being driven by short term profits | | |
| **1921–60** Research taking place was closely linked to business. R&D department was established at each plant | **1921** – Research section set up | |
| | | **1928** – Sintered alloy business unit is set up to produce dices used for elongating wires |
| | | **1946** – R&D Unit in Itami (near Osaka) established |
| **1950s** – Begins research on milliwave communications | | |
| **1956–65** President Kitagawa stresses the importance of developing technological | | |

| *Critical events* | *Organizational features* | *Role of other agents* |
| --- | --- | --- |

capability in-house
rather than licensing
foreign technology

**1960s** – Top
mamagement adopts
an aggressive
diversification strategy
in order to increase
sales of products other
than cables and wires.
Aims to reduce the
proportion of sales
derived from cables
and wires to 50% in
twenty years time
**1960s** – Diversification
either based on same
core technologies or
markets
**1960s** – Enters brake
business. Brakes were
used in the production
of wires
**1961** – Begins research
on III–V
semiconductors.
Researchers in the
Metals division
wanted to see what
kind of a device could
be made with gallium
arsenide which was
much faster than
silicon

**1959** – R&D centre in
Osaka established to
work on metals such
as copper wires and
other materials and
electricity
**1959**–Gum, plastic
business unit set up
**1960** – R&D centre in
Itami completed to do
research on III-V
semiconductors,
energy and
automobile related
technologies
**1960** – R&D divided
into main areas of
Heavy Electrical,
Weak Electrical
(Communications),
Metals, other
Materials and
chemistry

**1964** – R&D group in
Yokohama (near
Tokyo) established to
do research on
telecommunications

| *Critical events* | *Organizational features* | *Role of other agents* |
| --- | --- | --- |
| | | **1966** – Kao at STL presents a paper predicting the coming era of optical communications |
| **1967** – Research on optical fibres starts; Initially 7–8 people working on the project<br>**1969** – Conducts trial and error experiments with properties of glass[1]<br>**1969** – Research on optical fibres moves to the stage of applied research to work on mass production techniques<br>**1970** – Following NTT's move, terminates research on milliwave communications and transfers half of the people to work on optical fibres and the other half to semiconductor devices | **1970** – Semiconductor research unit becomes Electronic Device Material Development unit (an intermediate stage before it becomes a division)<br>**1970s** – Research on weak electricity, such as electronics used in transportation systems for traffic control, automobiles, railways carried out<br>**mid 1970s** – Acquires the business unit of III-V semiconductors of a medium sized firm called Nikkei Kakou, hiring 5–6 specialists on epitaxial growth[2] | **1970** – Discovery of room temperature semiconductor lasers at Bell Labs<br>**1970** – Corning announces that it succeeded in developing low loss optical fibres of 20 decibel/km loss<br>**1970** – NTT puts a halt on research on milliwave communications and decides to see the progress in optical communications<br>**1970s** – Competing cable and wire firms, Furukawa and Fujikura also diversify into III-V semiconductors until **the early 1970s** – The largest customers are NTT followed by electric power companies |

| *Critical events* | *Organizational features* | *Role of other agents* |
|---|---|---|
| **1971** – Begins commercializing III-V semiconductors for use in gallium phosphide LEDs<br>**1971** – Long term goal outlined to double sales to 360 billion yen by 1975<br>**1972** – Cable related sales account for 70% of sales<br>**1973** – A pilot plant to produce optical fibres is built in Yokohama | **1971** – A major reorganization of R&D activities takes place, establishing an R&D headquarter, to conduct research cross cutting the company. R&D planning unit set up[3] | |
| | | **1973** – Receives a big order from a US firm for Gallium Arsenide Phosphide for use in Light Emitting Diodes for calculators<br>**1974** – Bell Labs announces that it succeeds in dveloping low loss optical fibres by MCVD |
| | **1975** – The number of people working on optical fibres reach 100 | **1975–80** Joint research with NTT, Furukawa and Fujikura takes place to develop Japanese technology (VAD) to produce preforms for low loss optical fibres |
| **1976** – Succeeds in producing optical fibres by plasma CVD method<br>**1976** – Begins sales of GaAs wafers | **1976** – Merges the research activities taking place on communication cables in different units | **1976** – Supplies optical fibres to Tokyo Electric Power, Kansai Electric Power which try out optical fibre network in their plants |
| **1977** – Begins production of optical fibres for supplying NTT<br>**1977** – Begins planning of production of ICs since they are | | **1977** – Completes installing a marine optical fibre communications cable attached to an electric power cable for the first time at a Japanese oil refinery plant. This |

| *Critical events* | *Organizational features* | *Role of other agents* |
|---|---|---|
| essential for optical communictions<br>**1977** – Enters the field of medical electronics systems | | type of product is called OPGW (Optical Fibre Composite Aerial Ground Wiring)<br>**1978** – Sponsored by MITI and the Ministry of Post and Telecom, HIOVIS (Highly Interactive Optical Visual Information System) trial takes place in Nara linking 158 homes in an optical network<br>**1979–82** Supplies optical fibre network totalling 8000 km to Argentine PTT<br>**1979–86** In the MITI funded National optoelectronics project, carried out work on component generic technologies for developing OEIC |
| **1980s** – Top Management actively promotes the corporate identity of 'Optopia', a future society made more prosperous by optical comm. and systems[4]<br>**1980s** – A decade of VICC, Vertical Integration, Computer and Communications, focusing on developing components and devices which were the missing link between the materials and | **1980s** – Venture business system becomes well established allowing researchers to commercialize their work | |

| *Critical events* | *Organizational features* | *Role of other agents* |
| --- | --- | --- |

systems for which they had expertise. Strategy shifts from being a materials supplier to a systems integrator and a merchant maker of components
**1980** – Develops image fibre, a bundle of some 30000 optical fibres to transmit image data
**1980s** – Becomes a leading maker of gallium arsenide wafers holding 50% market share in Japan

**1981** – Optoelectronics Development Business Unit established

**1982** – Optical fibre plant completed in Yokohama
**1983** – Commercializes engineering work stations
**1983** – Begins selling SumiLink, a Local Area Network used in office buildings, factory automation and plant control

**1983** – Establishes Sumitomo Electric Research Triangle to do research and production of optical fibres, in North Carolina of the US
**1983** – Number of people working on optical fibres reach 150. Holds 35% market share in the domestic market for optical fibres

**1984** – Supplies optical fibres to Tohoku Electric Power

**1985** – Plans to invest 340 billion yen on R&D from 1985 to

**1985** – Establishes Osaka Basic Technology

| Critical events | Organizational features | Role of other agents |
| --- | --- | --- |
| 1990. Emphasizes the following as growth areas: optoelectronics, new materials, system products such as engineering work stations **1985** – Sales of non cable and wires reach 50% of total turnover as originally planned 20 years ago | Laboratory to conduct research on new materials and systems and software **1985** – Strengthens R&D organization by putting Osaka laboratory, Itami laboratory, Yokohama laboratory under the R&D headquarters[5] **1986–91** The number of researchers working on optoelectronics grows from about 75 to 150 (40 working on systems, 110 on devices) **1987** – Activities of optoelectronics research were merged bringing together the research done on III-V semiconductors in Osaka and Systems in Yokohama and established as an Optoelectronics Research Lab in Yokohama | **Since 1987** – Automobile sector becomes the biggest client (reaching about a third of sales) followed by electric power companies and NTT **1991** – Seven competing cable and wire firms including Sumitomo Electric form a joint R&D laboratory to conduct research on environmental related issues |

[1] M. Miyazaki (1985).

[2] M. Miyazaki (1985).

[3] Japanese Management Association (1987).

[4] *Annual Report*, Sumitomo Electric (1989).

[5] Japanese Management Association (1987).

# Appendix E.4   Siemens

| Critical events | Organizational features | Role of other agents |
|---|---|---|
| **1847** – Established by two entrepreneurs Siemens and Halske who want to start a venture on communications equipment | | |
| | **1850** – From its earliest days, Siemens actively forms alliances with other firms, a characteristic feature which remains to present | |
| **1871** – Constructs the Indo-European telegraph line, a distance of 11,000 km, enabling Siemens to be considered as the leading marine cable producer in the world[1] | | |
| | **1891** – Philips NV established to make lamps<br>**1896** – Discovery of X rays<br>**1907** – German PTT asks for Siemens' help to carry out their large scale plans to convert telephones to aumomatic working<br>**1912–1918** Enters into an arrangement with British rivals B.T.H and GEC to exchange technical information and pool patents of filament lamps | |

| Critical events | Organizational features | Role of other agents |
|---|---|---|

**1914–19** During World
War I, Siemens
engaged in production
of military equipment

**1914** – Siemens has
presence in forty nine
countries, with a
workforce of over
80,000 people

**1919–30s**
Expands capacity,
increases turnover and
profits.
Becomes involved in
power generation
equipment, electrically
driven railways and
electric lighting
systems. Diversifies
overseas with the
founding of
subsidiaries in China,
Indonesia, Mexico,
Brazil, Switzerland,
Czechoslavakia, often
through acquisition of
local firms.
Diversifies product
range to cover medical
electronics, and other
products in electrical
engineering[2]

**1919** – The first
corporate research lab
'Research lab of the
Siemens Work' is
completed after the
end of the war
**1919** – Osram
company comes into
existence, comprising
Siemens, the AEG,
and the Auer
companies with their
respective lamp
factories

**1920s** – Telefunken is
formed through a joint
venture with AEG
**1921** – A semi-public
undertaking formed,
known as the German
Fernkabelgesellschaft
of which German Post
Office, Siemens, AEG
and another company
are partners.
Undertakes the supply
and laying of long
distance cables. The

| *Critical events* | *Organizational features* | *Role of other agents* |
|---|---|---|
| | | membership is extended to four cable makers<br>**1923** – Enters the Japanese market with the formation of the subsidiary company Fuji Electric, by concluding an agreement with Furukawa Electric |
| | **1929–33** During the world recession, Siemens reduces workforce from 38900 to 18600 in Germany[3] | |
| | **1935–45** Expansion of the Central Research laboratory for communication engineering takes place | **1935** – Forms a joint venture with Fuji Electric in Japan to set up a communications equipment company which later becomes known as Fujitsu |
| **1945** – After the second world war, about 80% of capital assets were lost. Begins the process of rebuilding ruined factories and construction of new ones | | |
| **1950s** – Enters many diverse fields such as data systems and semiconductors | | **1950–6** Siemens' EMD switches used instead of cordpairs by West German PTT |
| **1951** – Scientists begin work on III-V semiconductors. The researchers had already worked on silicon and germanium detectors | **1951** – Research laboratory of Siemens Schukertwerke refounded to do research on application of automation to industrial power system and equipment, and basic research in the fields | |

| *Critical events* | *Organizational features* | *Role of other agents* |
|---|---|---|
| | of physics and chemistry, materials analysis and data processing | |
| | | **1955** – The nation wide DDD 'Direct Distance Dialling' program of the West German Post begins and Siemens benefits by supplying switching equipment |
| **1956** – Enters the field of nuclear engineering<br>**1957** – Siemens comes up with the first prototype all transistor computer<br>**1960s** – Scientists begin work on ruby lasers<br>**1963** – Scientists begin research on semiconductor lasers<br>**1964** – Develops solid state laser for industrial application<br>**1965** – Launches system 4004, a hierarchical family of data processors | | **1965** – Constructs and installs the first automatic mail handling system for the West German Post |
| | **1966** – Previously separated 'weak current' and 'strong current' related companies, Siemens Halske, Siemens Schuckertwerke and Siemens-Reiniger-Werke were merged in Siemens AG<br>**1966** – Laboratory for Electroacoustics in Karlsruhe which had | **1966** – Satellite communication begins |

| Critical events | Organizational features | Role of other agents |
|---|---|---|
| | been under construction for 14 months is put into service | |
| **By 1967** invested about 75 million DM on nuclear engineering | **1967** – Some reorganization of semiconductor laser research. Transfers people from semiconductor lasers to fuel cell research | |
| **1967** – Commercializes micromachining solid state laser which is a highly versatile machine tool | | |
| **1968** – Research on silicon transistors for use in colour TV sets | | |
| | **1969** – Reflecting the merger in 1966, a major reorganization of R&D takes place and a corporate R&D unit is founded as one out of five corporate functions | **1969** – Forms a joint venture in the area of nuclear technology with AEG which becomes known as Kraftwerk Union. |
| **1970s** – Siemens is first in the world to produce Gallium Arsenide Microwave Integrated Circuits, components for satellite TV | | **1970s** – Fujitsu begins OEM supply of its M380 series, which account for 10% of Siemens' computer sales |
| **1971** – Soon after the discovery of semiconductor lasers, Siemens researchers are able to achieve the same. However, the semiconductor lasers had to be cooled at cryogenic temperatures with liquid helium, and researchers stopped working on them, and began to work on red light emitting diodes, followed by infrared light emitting diodes. | | |

| *Critical events* | *Organizational features* | *Role of other agents* |
|---|---|---|
| LED arrays also developed for pumping gas lasers used in materials processing **1976** – Restarts research on semiconductor lasers | | |
| | **1977** – Spends about 400–500 Million DM a year on R&D | |
| **1980** – Objective of Siemens is to manufacture electrical/electronic goods and systems in advanced technology sectors and higher margin sectors in mature industries such as power engineering, thereby generating above average growth rates.[4] Debate was held internally whether to enter consumer electronics market but decided against it | **1980s** – Communications and Systems group established **1980s** – Siemens divided principally into seven groups including KWU, the Nuclear power plant subsidiary, Telecomm. Semiconductors and Components group, Comm. and Information Systems group, Power Engineering group, Installations group, Medical Engineering group. It is highly decentralized, with each group adopting different management styles. Each group director is on the management board and fully accountable for its performance. Five corporate divisions, including finance, personnel, Corporate Planning, R&D planning, Corporate Sales and Marketing exist | **until 1980s** – Siemens maintains close ties with the German government and PTT selling telecomm. equipment and other electrical and power equipment **1980s** – Forms alliances with Intel, National Semiconductor, AMD in the US |

| *Critical events* | *Organizational features* | *Role of other agents* |
| --- | --- | --- |

**1981** – Switches to InGaAsP semiconductor lasers in the 1.3 μm region

**1984** – Top management realizes that Siemens' markets have changed and that they can no longer continue to rely on lucrative Government contracts. Designates Office Automation, Telecommunications and Semiconductors, Computer Integrated Manufacturing as four key areas[5]

**1984** – Joins with Philips in a $2 billion project to produce 1 Mega bit DRAM chips

**1985** – Buys from Toshiba technology to produce 1 Mega bit CMOS chips

**1985** – Hands over to the German PTT wavelength division multiplexing optical communication system at the 1500 nm wavelength

**1986** – Siemens' share of the world's public network is around 6% which puts the firm in third place behind ATT and ITT[6]

**1986** – A huge R&D complex built in Munich for around DM1.2 bn whose main emphasis lies in microelectronics, IT and manufacturing technology. 30000 employees engaged in R&D worldwide. 97% of R&D funded internally, 3% from external sources

**1987** – Begins production of 1 M DRAM chips

| *Critical events* | *Organizational features* | *Role of other agents* |
| --- | --- | --- |
| **1988** – Half of Siemens' sales comes from Germany and 25% from the rest of Europe, US accounting for only 10%. Top management goal to increase US sales until those revenues match European revenues. Agrees to take over from IBM all the manufacturing part, and half the sales part of Rolm, an American company specializing in PBXs | **1988** – Another major reorganization of R&D takes place, throwing off its highly centralized management structure for small independent business units. Company reorganized into 5 groups (fifteen business units each with sales and marketing staffs) and management functions streamlined, reassigning 9000 employees at headquarters. The new organization designed to increase flexibility, impact and competitiveness.[7] Corporate R&D divided into two groups of Materials science and electronics; Applied computer science and software. Optoelectronics research conducted in the former group in the three departments of photonics, active devices, epitaxy; R&D committee consists of the directors of the 15 business units, Core Technology Officers, directors of the R&D lab, R&D planning who meet 4–6 times a year | **1988** – Ties up with GEC to launch a hostile bid for Plessey <br> **1988** – Siemens, Philips and SGS-Thomson agree to cooperate in developing 16 Mega bit chips through the $4 billion JESSI project |
| **1990** – Acquires Nixdorf, a leading computer company in Germany | | |

| Critical events | Organizational features | Role of other agents |
|---|---|---|
| | | **1991** – Deregulation of German PTT takes place |

[1]Siemens (1957).
[2]Siemens (1957).
[3]Siemens (1957).
[4]Vickers da Costa (January 1986).
[5]*Business Week*, 'Siemens Speeds Up' (20 February 1989), p.17.
[6]Vickers da Costa (January 1986).
[7]*Electronic Business*, 'Siemens Restructures R&D Closer to End Markets' (20 March 1989).

# Appendix E.5 Summary of GEC

| Critical events | Organizational features | Role of other agents |
| --- | --- | --- |
| **1886** – GEC formed. First products were bells and other home fittings | | |
| **1890** – GEC becomes a public company operating under the slogan 'Everything electrical' | | |
| **1909** – Production of OSRAM lamps begins | | |
| **1914–17** During the first world war, GEC transforms itself into a major industrial company by making important contributions to the war effort | | **1914–17** The government becomes the most important customer, accounting for 90% of production |
| **1921** – A subsidiary Hotpoint Electrical Appliances set up | **1921** – GEC Research Laboratories established at Wembley | |
| **1945** – Becomes the most consumer oriented company in the industry, prospering by meeting the postwar boom. Pursues a policy of making all kinds of electrical goods | | |
| | | **1950s–70s** Heavy dependence on the public sectors and government contracts, such as defence and the Post Office. GEC also holds commanding position |

245

| Critical events | Organizational features | Role of other agents |
|---|---|---|
| | | in diesel engines, gas turbines and other power engineering equipment. The company benefits from old British ties to places like Hong Kong and Africa |
| **1950s** – Enters the field of optoelectronics, III-V semiconductors carried out at Hirst Research, LEDs for display applications at Marconi Research<br>**1950s–70s** GEC builds competence in civil and military communcations such as public, radio and microwave communications | | |
| | | **1956** – Joint Electronic Research Committee with the Post Office set up |
| | **1957** – Rushton Research Centre opened<br>**1957** – At Plessey, the first factory of transistors begins production<br>**1960s** – More entrepreneurial freedom is given to the heads of manufacturing divisions subject to financial criteria | **1960s** – Central Electricity Generating Board (CEGB) embarks on an ambitious programme of power station construction, reacting to the power shortages of the fifties. Its forecasts were overoptimistic, and results in massive cut back in the late 1960s. |

| *Critical events* | *Organizational features* | *Role of other agents* |
|---|---|---|
| | | Plessey attempts to take over English Electric but fails, precipitating the GEC – English Electric Alliance |
| **1962** – Begins research on gallium arsenide **1963** – Lord Weinstock takes control and transforms GEC from being a large, ailing company into an efficient industrial group, by implementing strict financial control and closing down loss making units, cutting down the central office and delegating responsibility to operating executives. GEC's strategy to create employee allegiance to the operating company in which they work[1] **1967** – GEC acquires Associated Electrical Industries | | |
| **1968** – Takes over English Electric which had absorbed Marconi and Elliot Automation. GEC becomes a company of range and size that can compete on a truly international scale | **1968–70** After the mergers with AEI, English Electric, reduces the work force of the three companies by between 15 000 and 20 000, or close to 10% of the total (for example closes AEI's telecomm. factory at Woolwich) Overheads are reduced, manufacturing capacity | |

| *Critical events* | *Organizational features* | *Role of other agents* |
|---|---|---|
| | concentrated, R&D pooled and sales effort intensified | |
| | **Up Until 1970** – Few unifying technological linkages between the constituent firms which comprised GEC[4] | |
| **1970s** – GEC falls behind in mainstream semiconductor technology[2] | **1970s** – Plessey's semiconductor business grows steadily, becoming the largest semiconductor manufacturer in Britain. GEC is the fourth | **1970s–80s** Largest customers include MoD, Post Office, British Gas and British Rail. MoD contracts are risk free, since GEC can charge them on a cost plus basis |
| **1971** – Shuts down Marconi Elliot Automation which is making chips shortly after acquiring the company[3] | | **1971–81** Users of GEC's computer series include Post Office, CEGB, MoD, British Steel, Science and Engineering Research Council, universities and so on, who are GEC's main customers in other fields |
| **1972–1982** Turnover of GEC grows sixfold to £5.46 billion | | |
| | | **1977** – Agreement signed between GEC, STC, Plessey to develop System X for BT |
| **1979** – A marked expansion of resources in System X | **1979** – Derek Roberts is hired to become the Director of Research. Since then, the small 25 financial corporate units begin to be pulled together with increased central coordination[5]. Centrally directed | **1979** – Forms a joint venture with Hitachi to make TV receivers and music centres in the UK |
| **1979** – Forms a deal with Fairchild to produce micro-processors and memories based on MOS technology, but the deal is cancelled after | | |

| *Critical events* | *Organizational features* | *Role of other agents* |
| --- | --- | --- |

Fairchild is taken over by Schlumberger. GEC's chips are made at Marconi Electronic Devices. More than two thirds of the chips are sold to other GEC companies and most can only be used in radars or small volume applications . GEC does not have CMOS capability[6]

**1980** – Acquires A. B. Dick, a Chicago based company that makes office equipment
**1980s(Plessey)** – Sales of semiconductor chips grows from $20m in 1982 to over $100m in 1987, possessing leading technology in some areas[7]

research which previously had been negligible comes to occupy a quarter of the effort of GEC's central research organization at the Hirst Research Centre
**1979** – The workforce decreased to 184 000 from 234 000 in 1970, having closed down or sold a number of operations

**1980s** – Relies on Government subsidy for investments in semiconductors
**1980s** – Active collaboration between GEC and Plessey for the Alvey project
**1980s** – Increasing reliance on other firms' technologies in telecom. For example sells PBX licensed from BT and Northen Telecom[8]
**1980s** – GEC is a keen participant in collaborative projects, such as ESPRIT and Alvey[9]
**mid 1980s** – Close relationship with big customers ending due to deregulation and changing procurement policies.[10]

**1981** – Acquires Picker, a US maker of medical instruments

**1981** – Organization consists of Electronics, Automation and Telecomm.; Power Engineering; Components; Cables

| *Critical events* | *Organizational features* | *Role of other agents* |
|---|---|---|
| | and Wires; Industrial and Consumer products | |
| | **1981** – Sir Richard Clayton moves from the Navy | |
| | **1982** – Hires Cyril Hilsum who had done pioneering work on liquid crystal displays at RSRE | **1982** – BT selects Plessey as the primary contractor and GEC as the secondary contractor for System X[11] |
| | **1982** – Sets up a new office of the future subdiary, GEC Information Systems, bringing together parts of GEC's companies, GEC Computers, GEC Viewdata, GEC Telecom, and Reliance Systems. A.B.Dick remains separate | |
| | **1983** – Of the $55 million spent on R&D at the main research laboratories, roughly half is funded by the operating companies, with the remainder equally divided by the company's own funds and externally commissioned research | **1983** – Of the Electronics, Automation and Telecomm. group 60% of sales goes to the MoD |
| | **1983–4** Former Thatcher government's foreign minister Lord Carrington is chairman of GEC, followed by James Prior | |

| *Critical events* | *Organizational features* | *Role of other agents* |
|---|---|---|
| **1984** – Acquires Yarrow Shipbuilders<br>**1984** – System X is brought into production<br>**1984** – Attempts to bid for British Aerospace, but fails<br>**1985** – Attempts to bid for Plessey, but is blocked by the UK government since a merger would lead to a major loss of competition in the supply of defence electronic equipment[12] | **1984** – Sets up Research Advisory Group to assess the research done within the three major research centres; Cooperation within GEC encouraged<br>**1985** – GEC Research, a holding company for GEC's five research centres is established. Of the £60m earnings, 47% comes from GEC units, 35% from the Goverment and the rest from corporate funds.[13] GEC Research is to undertake basic work on new materials, device structures and systems that are likely to become important in the 1990s<br>**1985** – Existing eight divisions are the same as of 1979 | **1984** – BT invites tenders from suppliers marking an end to the 'telecom club' and selects Thorn Ericsson to supply SYSTEM Y<br><br>**1985** – Thatcher government criticizes GEC of its reliance on government contracts, of shedding its workforce while increasing its cash mountain and for failing to invest in new technologies[14] |
| **1986** – Announces the formation of GEC Automation Projects, an amalgamation of GEC Electrical Products and several other companies as a consequence of technological convergence<br>**1986** – Celebrates centenary years. A major exhibition held in Wembley attended by the Queen<br>**1986** – Weinstock sets up a 25 member board of management on the | **1986** – With the $1.5 billion cash surplus GEC Finance is created<br>**1986** – Of the £650 million spent on R&D, about 10% spent on the work undertaken by 2400 scientific research staff at the three research centres, Hirst Research Laboratory, Marconi Research Centre, and Engineering Research Centre. Alvey support comes to £3 | **1986** – GEC collaborates with Thomson-CSF and AEG competing for the development and production of a new weapon locating radar for Western Europe<br>**1986** – As part of ESPRIT project, GEC successfully develops A4 size LCDs based on polycrystalline and amorphous silicon TFTs<br>**1986** – MoD cancels NIMROD, the |

| Critical events | Organizational features | Role of other agents |
| --- | --- | --- |
| West German model to revitalize the company | million. Work is done to expand its expertise in key enabling technologies.<br>**1986** – GEC consists of some 160 operating companies | airborne early warning system for which GEC fails to meet specifications for nine years.[15]<br>**1986** – US DOD awards a $100m contract to GEC Avionics<br>**1986** – Thatcher government refuses to underwrite an export subsidy for a System X sale to India<br>**1986–88** GEC takes part in the JOERS scheme, Joint Opto-Electronics Research Scheme, and works on Optically powered distributed sensor network<br>**1986** – Receives a contract worth £400m from the MoD for a torpedo<br>**1986** – GEC, Siemens, Thomson, and Philips participate in EUREKA project, a pan-European collaborative R&D project |
| **1987** – Disposes the optical fibre business<br>**1987** – Acquires Creda group, Gilbarco and Lear Astronics<br>**1987** – GEC is a key participant in the multi-national consortium to develop the defensive aids for the European Fighter aircraft | **1987** – Some 1000 employees are made redundant as a consequence of the cancellation by the MoD of the NIMROD project<br>**1987** – Marconi Underwater Systems establishes a new Maritime Research Unit at Cambridge to investigate advanced | **1987** – Forms a joint venture with Philips to combine their medical systems activities with GEC's subsidiary Picker International<br>**1987** – Withdraws from CMOS R&D supported by the Alvey programme |

| Critical events | Organizational features | Role of other agents |
|---|---|---|
| **1987** – Withdraws from producing semiconductor lasers when other producers products become cheaper | acoustics and signal processing. The £18m Advanced Technology Centre is also opened. | |
| | **1988** –The telecomm. businesses of GEC and Plessey are merged into GPT (GEC Plessey Telecom) the former having a 60% interest, the latter 40% interest | **1988** – Defence still the highest earner, followed by telecom, power generation and electrical equipment **1988** – MoD's Levene opposes 'cost plus' contracts |
| **1989** – GEC ties up with Siemens to bid for Plessey. GEC regains R&D expertise in semiconductor lasers which is done at Caswell | | |

[1] *Wall Street Journal*, 'Weinstock's Way' (3 October 1983).

[2] Morgan *et al.*, CICT paper (January 1989).

[3] *Electronics Weekly* (14 January 1987), p.16.

[4] Memo on interview with D.Roberts by M.McLean (SPRU) (27 June 1985).

[5] Memo on GEC by M. McLean (SPRU) (1979).

[6] *Computing the Magazine* 'A Heavy Weight Hogs the Market' (26 September 1985).

[7] Memo by M. Hobday (SPRU) (28 November 1988).

[8] *Sunday Times* (9 October 1985), 'GEC – How Weinstock's Giant lost its vital spark'.

[9] Morgan *et al.* (January 1989).

[10] *Sunday Times* (9 October 1985), 'GEC – How Weinstock's Giant lost its vital spark'.

[11] Morgan *et al.* (January 1989).

[12] CICT Working paper no.2 (1989), 'The GEC-Siemens bid for Plessey'

[13] Memo by M. McLean (27 June 1985).

[14] House of Commons (28 April 1985).

[15] *Management Today*, 'How Bad is GEC?' (January 1989), p.42.

# Appendix E.6   Summary of STC

| Critical events | Organizational features | Role of other agents |
|---|---|---|
| **1883** – Western Electric of the US opens an office in London | | |
| | | **1920s–80s** Heavy dependence on the Post Office (BT) |
| **1925** – When Western Electric is acquired by ITT, the British subsidiary is renamed Standard Telephones and Cables Limited | **1925–70s** STC is run at arm's length by ITT, which allows it a large degree of autonomy within the terms of its general corporate strategy and financial disciplines. STC is divisionalized from its early days. R&D locations are scattered within the company until STL is established | |
| **1930s** – Alex Reeves of STC invents Pulse Code Modulation. Develops and installs the first 12 MHz system in the world. Develops and installs the first 4 MHz system in Europe operating over coaxial cable on long distance trunk comm. routes | | |
| | **1950s** – Standard Telephone Laboratory established conferring greater dignitiy and self importance | |
| | | **1956** – Joint Electronic Research Committee with the Post Office |

| Critical events | Organizational features | Role of other agents |
|---|---|---|
| | | set up (disbanded in 1968) |
| | | **1950s–60s** Supplier to British Post Office for submarine cables, coaxial cables |
| **1960s** – Becomes a world leader in the field of submarine comm. systems | | **1960s** – Microwave technology promoted by Bell Labs, STC sought an alternative and hence developed optical comm. backed by British Telecom |
| **1962** – 5–6 people begin to work on semiconductor lasers, its earliest application is for military | | |
| **1962–72** Work carried out on LED for applications such as ship to ship comm., illumination for night vision | | |
| **1966** – C. Kao and G. Hockham, two STL researchers publish the first proposal on optical fibre communications | | |
| **Early 1970s-** Research on liquid crystal displays begins. Their LCDs used as display panels at Heathrow airport | | **1970s** – STC has been a major contributor to BT's programme to introduce optical transmissions network |
| | | **1970s** – Dr. O'Hara from BT, who is a keen promoter of single mode optical comm. decides early on that single mode is the way to go |
| **1972** – Semiconductor lasers put into production | | |

| Critical events | Organizational features | Role of other agents |
|---|---|---|
| | **1973** – A design centre for chips established | |
| | | **1977** – Work starts on the first TXE4A, Britain's most advanced telephone exchange to be supplied to the Post Office in 1980. TXE4A is a cost reduced, technology updated version of STC's TXE4 exchange |
| | **1978** – At this point in time, the following business groups exist:- Switching, Land Line Transmission Systems, Comm. Cable, Optical Comm. Systems, Submarine Comm. Systems, Electronic Equipment, Subscriber Apparatus, Project Management | **1978** – The Government gives ITT £2m for investing in expanding its memory production plant in Kent, to produce advanced chips in particular 64K DRAMs and MOS techniques[1] |
| **1979** – STC becomes a public company | | **1979** – British Telecom Systems formed by GEC, Plessey, STC and BT to oversee the development of System X, an advanced PCM based switching and transmission system which BT plans to implement in the 1980s |
| | | **1979** – ITT decides to offer 15% of STC for sale[2] |
| **1980** – STC completes installation of world's first deep water optical fibre based undersea telephone cable. A trial 9.5 km loop of cable was laid | | **1980** – ITT's anti trust suit against ATT, accusing the company of refusing to buy equipment from outside suppliers reaches an agreement |

| Critical events | Organizational features | Role of other agents |
| --- | --- | --- |
| by the British Post Office cable ship on the west coast of Scotland[3] | | by agreeing to purchase as much as $2 billion in products and services from ITT. Since STC is busy with System X and does not make the ITT 1240 exchanges it is not affected[7] |
| **1981** – London's first optical comm. link in the public telephone network was officially handed over to BT<br>**1982** – STC has to withdraw from System X project to develop switching equipment and decides to concentrate on transmission and terminal equipment based on the perception that switching will be increasingly per-formed by chips inside the terminals[4]<br>**1983** – Freed from the ownership of ITT, STC embarks on a restructuring process.[5] Also the process of acquisitions begins<br>**1983** – £60 million spent on upgrading the design and production capabilities of semiconductors and ITT Semiconductors purchased.[6] Believes that its future success | **1983** – The withdrawal from System X results in a loss of 400 jobs<br>**1983** – Nigel Horne, former managing director of GEC Informations Systems joins STC and restructures the company into six groups of telecomm., business systems, international comm., switching and | **1981** – To date, over 40% of optical fibre systems for the BT network have been ordered from STC[8]<br>**1982** – ITT decides to cut down its 75% majority holding in STC down to 35%. The link between the two companies is to continue in the form of an agreement to share technologies, giving STC access to ITT's R&D network[9]<br>**1983** – Buys IAL the comm. and airport service company which has strong pres-ence in the US and the Middle East from British Airways to counteract the weak-ened position of STC in overseas markets after the divestiture by ITT |

| *Critical events* | *Organizational features* | *Role of other agents* |
| --- | --- | --- |
| depends on its ability to make and sell semiconductors. STC has 60% of the submarine cable market | transmission, marine cables and components. Three groups become limited companies, they are STC Telecomm. Limited (Switching and Transmission), STC Business Systems Limited, and STC Comm. International Limited (marine cables)[10] **1983** – STC building optical fibre production capacity (currently at 25000 km a year, building towards 50000km)[11] **1983** – STC Technology Ltd. formed to offer high tech service to government and industry as well as STC's units. Facilities include STL and IDEC, a specialized systems design house. Combines the resources of 1500 specialists[12] | **1983** – Sales to BT account for 33.2% of total turnover[13] **1983** – Wins a $100 million defence comm. contract for the British forces in Germany |
| **1984** – STC acquires ICL in order to achieve convergence of computers and comm. Fujitsu also takes part in the joint venture **1984** – Work on visible wavelength Light Emitting Diodes terminated. Shift resources to semiconductor lasers **1985** – STC runs into financial crisis | **1984** – STC Business Systems division transferred to ICL **1984** – Advanced manufacturing centre and Production Technology Centre formed **1985** – Arthur Walsh of GEC joins STC as | **1984** – Wins a contract worth over £100 million for an undersea telecomm. cable linking Australia, Singapore and Indonesia |

| *Critical events* | *Organizational features* | *Role of other agents* |
| --- | --- | --- |

**1985** – STC wins award for smectic type liquid crystal displays which have excellent contrast, good viewing angle and good contrast

**1985** – Decides not to make investments to mass produce liquid crystal displays and exits. The researchers continue to work on liquid crystals in spatial light modulator applications

**1985** – Outlines its long term plans to become a significant player in the global information technology market

**1985** – STC decides to concentrate making only those key components which would give added value to the company. Up to six component businesses are sold or closed.[14]

chief executive and is involved with firefighting. He institutes strict financial control methods using GEC style techniques. Walsh and Lord Keith cut back on central administration and push operating responsibility down to the divisions. Staff reduced by 8400, and peripheral businesses eliminated

**1985** – Forms a 1000 strong Networks division bringing together people from the IAL Data Comm division, Network Business Centres, the PABX Business Unit in Foots Cray, STC's IDEC in Stevenage a software house acquired in 1983, and ICL International Network Services.

**1985** – As part of a cost reduction programme, Components activities restructured transferring central headquarters functions including personnel, marketing to the manufacturing divisions

**1986** – STC has taken 8400 out of its total headcount leaving 43200 at year end

**1986** – Signs a £6.5 million ESPRIT contract to research methods on building knowledge based systems, in

| *Critical events* | *Organizational features* | *Role of other agents* |
|---|---|---|
| | | partnership with Scicon, French Cap Gemini Sogeti, and SCS in Hamburg |
| **1987** – Spending on optoelectronic-components increased | | **1987** – Collaborates with LSI Logic in the US on ASICs |
| | | **1987** – Northern Telecom holds 27.5% of STC |
| **1990** – Decides to sell 80% of its holdings in ICL to Fujitsu, realizing that STC was not big enough to be a major player in both communications and computers | **1990** – About a third of the funding for R&D taking place at STL comes from the corporate headquarter, a third from the operating companies, and the rest from other sources such as government contracts and ESPRIT. A third of the corporate funded portion is allocated to optoelectronics | |
| **1990** – STC is acquired by Northern Telecom of Canada which held 27.5% stake since 1987 | | |
| **1991** – Disposes optical fibre division | | |

[1]*Financial Times* (21 December 1978).
[2]*Financial Times* (14 June 1979).
[3]STC News Release (February 1980).
[4]Memo by M. McLean, SPRU (6 August 1984).
[5]*The Times*, (14 April 1983).
[6]STC *Annual Report*, (1983).
[7]*Financial Times* (29 February 1980), p.28.
[8]STC News Release (1983).
[9]STC News Release (1983).
[10]STC News Release (February 1983).
[11]*The Times* (14 April 1983).
[12]STC News Release (February 1983).
[13]STC *Annual Report* (1983).
[14]*Electronics Times*, 'For sale signs go up at STC's Norwich base' (18 July, 1985).

# Appendix E.7 Summary of Sony

| Critical events | Organizational features | Role of other firms |
| --- | --- | --- |
| **1946** – Established by Ibuka and Morita; the charismatic entrepreneurial characters of the founders contribute to the company's growth | **1940s** – Total number of employees reach 50–60; In order to develop a tape recorder recruits physicits and chemists[3] | |
| **Late 1940s** – Decides to develop a tape recorder<br>**1949** – Prototype of tape and tape recorder made<br>**1952** – The founder of Sony goes to the US to sell tape recorders[1] | | **Late 1940s** – Receives an order for broadcasting equipment from NHK (Japan Broadcasting Corporation) and the Japanese government. Sold tape recorders initially to courts of justice, and then to schools<br>**1954** – Acquires a license from Western Electric to develop transistors |
| **1955** – Begins production of transistor based radios[2]<br>**1957** – Invents Ezaki diode for which Ezaki was later awarded a Nobel prize | **1957** – Establishes a semiconductor division | |

| Critical events | Organizational features | Role of other firms |
|---|---|---|
| **1957** – Produces world's first pocket size transistor based radio | | |
| | **1961** – Central Research Centre established | |
| **1963** – Launches transistorized small VTRs | | |
| | | **1965** – Forms a joint venture with Tektronis of US |
| **1968** – Trinitron colour TV launched **1969** – Begins research on MOCVD | | |
| | **1971** – Morita becomes president **1972** – Audio Technology Centre established | **1972** – Philips begins work on optical disks |
| **1974** – Begins research on optical disks, a decision made by top management **1975** – Begins research on semiconductor lasers using LPE **1978** – Component technology to develop semiconductor lasers switch to MOCVD **1979** – Chooses Sharp's semiconductor lasers having tested samples from Hitachi and Mitsubishi Electric as well **1979** – Walkman launched **1979** – Room temperature oscillation of semicon. lasers achieved | | **1978–9** Joint collaboration with Philips to develop optical disks **1979–85** Doesn't take part in MITI's optoelectronics project |

| Critical events | Organizational features | Role of other firms |
|---|---|---|
| **1980** – CCD video camera launched<br>**1980s** – It takes 7–8 years to develop CD-ROM | **1980s** – Monthly resesarch presentations held; Formal presentations held 7 times a year. Core technology champions exist<br>**late 1980s** – About 15 Tokken type projects in progress | **1980–1** Collaborates with Konika to develop aspheric lens for CD player<br>**1980s** – Appoints the director of NTT's research lab to head their research in comm. |
| **1982** – CD player launched | **1982** – Ooga becomes president and Morita chairman | |
| **1983** – Sales drops by 10%, operating profits by 52%<br>**1983–4** With the slowing down in sales, top management decides to increase sales of non-consumer related products (plans to bring the ratio to 50:50 from the current ratio of 1:3). Outlines AV&CCC Audio Visual, Computers, Components and Comm. strategy, putting emphasis on basic and applied research related to component core technologies.<br>The new president Ooga believes in the integration of voice, sound, audio, visual data etc. Decides to turn Sony from being a captive to a merchant maker of components<br>**1984** – Decides to switch production | **1983** – Establishes SBUs as profit centres, each being in charge of marketing, production and R&D (an exception is made of semiconductor and production technology divisions which are supported by the headquarter. However, even the latter is made a profit centre so it has to sell technology internally as well as externally)<br><br>**1984** – Establishes Component | |

| Critical events | Organizational features | Role of other firms |
|---|---|---|
| process of semicon. lasers from LPE to MOCVD (LPE had been used since 1981) **1984** – Internally made semiconductor lasers begin to be used in their CD players. Volume production begins | business unit. Sales growth from 30 billion yen in 1985 to 120 billion yen in 1988 | **1984** – Launches videodisk player with OEM products supplied by Pioneer |
| **1985** – Enters FA (factory automation) field such as assembly robots **1985** – Starts a project for developing a 32bit engineering workstation 'NEWS'. Speedy development in six months by using 'Venture' system **1987** – UNIX based NEWS launched and becomes the leading maker, sales reaching 10 billion yen in 1987. Creates Desk top Publishing market **1988** – Enters OA (Office Automation) field | **1985** – A new SBU organization starts delegating more power to the business units **1985** – FA unit set up within Production Engineering<br><br>**1987** – Super Micro business unit formed to sell engineering workstations<br><br>**1988** – Reorganization of R&D to complete the transition from being a product market driven R&D to a core technology driven system. Corporate Research laboratory established in which optoelectronics research laboratory employs about 100 people to do applied research on optical disks, optical comm., optical sensors and memories. Basic | **1985** – Launches in-house made video disk players by acquiring licence from Philips |

| *Critical events* | *Organizational features* | *Role of other firms* |
|---|---|---|
| | research on optoelectronic materials carried out in the Central Research Centre | |
| **1989** – Develops 4M SRAM chip; Announces plan to enter PC market | **1989**–16 SBUs exist, the largest one being the Audio group and the smallest one being the 'Super Micro' group | **1989** – Acquires Columbia Pictures **1989** – Reaches license agreement with MIPS Computer Systems for RISC (Reduced |

[1]Morita (1986).
[2]*Sony Annual Report* (1991).
[3]Morita (1986).

# Appendix E.8    Summary of Sharp

| Critical events | Organizational features | Role of other agents |
| --- | --- | --- |
| **1912** – Established as a pencil maker in Tokyo | **Early period** – The founder stresses the importance of making products that others would want to imitate<br>**1923** – Moves headquarter to Osaka after a major earthquake hits Tokyo | |
| **1925** – Succeeds in commercializing radios | | |
| | **1931** – Establishes R&D laboratory to conduct research on TVs<br>**1942** – Microwave Communications R&D section set up | |
| **1953** – Begins mass production of TVs, however does not produce the key components of TV tubes and relies on suppliers<br>**1960s** – Adopts a vertical integration strategy based on calculators[1]<br>**1960s** – Begins production of CMOS LSIs for use in their calculators | **1960** – Semiconductor R&D unit, circuit R&D unit founded to conduct R&D on the key technologies required for developing calculators | |
| | **1961** – R&D centre established to conduct research on calculators, solar | |

| Critical events | Organizational features | Role of other agents |
|---|---|---|
| | batteries and medical equipment | |
| **1962** – Begins volume production of microwave ovens<br>**1963** – Enters the field of III-V semiconductors (LEDs, semicon. lasers) to find substitutes for gas lasers which were bulky and unpractical to be used<br>**1963** – Succeeds in large scale production of single crystal solar cells (silicon photovoltaic cells) for use in Sharp's electronic calculators<br>**1966** – Launches world's first calculators using ICs<br>**1968** – Begins mass production of LEDs, and photodiodes for industrial equipment<br>**1970s** – Until this point, Sharp is an assembly maker whose main line of products centres on consumer electronics and AV products. Top management realizes that the market for consumer electronics would be saturated. Decides to put more stress on non consumer related field. Decides to develop the key components themselves | | |
| | **1970** – Builds a plant in Tenri to produce CMOS LSIs<br>**1970** – R&D centre transferred to Tenri, the current location | **1968** – RCA develops a prototype of Liquid Crystal Display<br><br>**1970s** – Sells broadcasting equipment to broadcasting companies |

| *Critical events* | *Organizational features* | *Role of other agents* |
| --- | --- | --- |
| **1973** – Commercializes world's first calculators using LCDs<br>**1974** – Begins research on TAB (Tape Automated Bonding) technology used in LCDs<br>**late 1970s** – Begins research on CCDs | | |
| | **1976** – Corporate Technological Planning division established to strengthen links between the Corporate Research Laboratory and the business units[2] | |
| **1977** – Develops a prototype of Japansese word processors<br>**1977** – President Saeki creates a system of Kinkyuu (Top priority urgent ) projects similar to Tokken which are monitored by the decision making group including the divisional and R&D directors, who meet every month<br>**1979** – Begins selling its first word processor at 2.95 million yen<br>**1979** – First company to succeed in developing VSIS type semiconductor lasers which were used in Sony's CD players | | **1977** – Applies to take part in MITI's optoelectronics project but gets rejected |

| Critical events | Organizational features | Role of other agents |
|---|---|---|
| **1980s** – Jump 80 Strategy implemented to bring the ratio of sales of consumer and non-consumer related products to 50:50, and to bring the ratio of domestic to overseas sales to 50:50[3] | **1980** – Begins setting up a 3 tier R&D organizational structure by forming applications oriented research laboratories in addition to the corporate level laboratories[4] <br> **1980** – Establishes Energy Research Laboratory | **1980–9** Contracts research from MITI funded research establishment NEDO (New Energy–Industrial Technology Development Organization) to develop amorphous solar batteries <br> **1980s** – Joint research on liquid crystal materials takes place with a supplier of materials <br> **late 1980s** – Hires someone from NTT to head research on comm. equipment |
| | **1981** – Establishes Production Engineering Laboratory and applications research laboratories on Semiconductors, Visual systems, Electronic equipment | |
| **1982** – Succeeds in breaking the 700 nm barrier and developing semiconductor lasers which emit light at 683 nm (suitable for use in HDTV and optical storage applications) | **1982** – Establishes Information System Research Laboratory to undertake research on (Office Automation), AI, and IT <br> **1982** – Establishes a laboratory to do research on Home Automation | **1982–92** Participates in MITI funded Fifth Generation Computer project ICOT |
| **1983** – Succeeds in volume production of Electroluminescent displays (which were used in NASA's space shuttle) | **1983** – Establishes a research centre in Tokyo to undertake research on core technologies such as new materials, VLSI, and new media | **1983** – OEM supply of telephones to US |

| Critical events | Organizational features | Role of other agents |
|---|---|---|
| **1983** – CMOS LSIs for driving LCDs developed<br>**Mid-1980s** – Begins work on high power lasers | **1983** – Establishes Audio System equipment research centre<br>**1983** – Establishes 'Living Soft Centre' to conduct research on changing consumers' needs and life styles<br>**1984** – Sets up IC business unit | |
| **from late 1980s** – Integration of hardware, software, systems and services emphasized.<br>Expands activities into DRAMs, SRAMs and memory chips | | |
| | **1985** – Establishes VLSI R&D Centre to work on mask ROM, SRAM, gate arrays, CCD and LCD drivers<br>**1985** – Establishes Communications Audio business unit to market telephones | **1985** – Privatization of NTT takes place, enabling Sharp to launch communications equipment such as telephones |
| **1986** – A prototype version of LC projection display developed<br>**1986** – Develops black/white DSTN (double supertwisted nematic) LCD<br>**1986** – System of Large Scale projects begins<br>**1987** – Puts on the market, electronic diary which gets awarded for the best product of the year | **1986** – LCD research centre set up<br><br>**1987** – Completes forming a three tier R&D organizational structure.<br>In addition to the seven corporate level | |

| Critical events | Organizational features | Role of other agents |
| --- | --- | --- |

**1987** – Launches magneto-optical disk system
**1987** – Begins volume production of colour LCDs

laboratories, there are the following:-
Precision engineering, Computer Systems, Production engineering, Energy, Generic Research centre in Tokyo

**1988** – Develops colour DSTN LCD (14 inches in size). Launches a flat panel TV (3 cm in thickness)
**1988** – Succeeds in developing a high power semiconductor laser emitting 150–315 mW, achieving the best performance in the world
**1988** – Notebook word processor launched

**1988** – Optoelectronics Device R&D laboratory is changed from being a corporate level research laboratory to an applications oriented research laboratory signifying its close links with the business units[5]

**1989** – LC projection display system put on the market, sells well in US and Japan
**1989–92** Invests 100 billion yen on capital expenditure on LCDs
**1989** – Invests additional 40 billion yen on building a plant to produce LCDs

**1989** – Establishes an Intellectual Property Centre

**1989** – Makes electronic diary technology freely available to third party software houses and other firms so that they can develop IC cards

**1990** – Liquid Crystal Division splits from Electronics Component business unit to become an independent business unit signifying its growing importance
**1990** – About 14 Kinkyu 'Top Priority'

| *Critical events* | *Organizational features* | *Role of other agents* |
|---|---|---|
| | projects in progress including HDTV **1990** – Establishes a basic research centre near Oxford to conduct research on optoelectronics and natural language processing | |

[1]Trigger (1990).
[2]Japanese Management Association (1987).
[3]Trigger (1990).
[4]Japanese Management Association (1987).
[5]Jojima (1990).

# Appendix E.9 Summary of Toshiba

| Critical events | Organizational features | Role of other firms |
|---|---|---|
| **1870** – Shibaura Electric founded<br>**1890** – Tokyo Electric founded. Begins producing light bulbs | | |
| | **1906** – Shibaura Electric establishes a R&D laboratory to conduct research on heavy electrical equipment. Decentralized and divisionalized. Gishichou (Technological Chief) System adopted | |
| **1933** – Succeeds in the first TV experiment in Japan | | |
| | | **1936** – Sells broadcasting equipment to NHK (Nippon Broadcasting Corp) |
| **1939** – Merger of Shibaura Electric and Tokyo Electric takes place to become Toshiba<br>**1942** – Japan's first radar developed<br>**1949** – Begins manufacturing black/white TV<br>**1952** – Begins development of transistors[1]<br>**1953** – Spins off consumer electronics | | |

| Critical events | Organizational features | Role of other firms |
|---|---|---|
| sales division as Toshiba Trading **1959** – Toshiba type helical scan VTR developed **1959** – Develops transistor based TV | | |
| **until the 1960s** – Primarily a consumer electronics maker deriving 60–70% of sales from this sector **1960s** – Researchers begin research on application of lasers in holography to form three dimensional images. Considered it was possible to use it as a large memory device for information processing.[2] Develops a holography equipment but too early for commercialization. The researchers then develop 'Tosfile' an optical disk filing system. Succeed in HeNe lasers. Begins research on semicon. lasers in the visible wavelength region | **1960** – Spins off Toshiba EMI | **1960s** – Cooperates with NEC to develop medium and large size computers |
| | **1961** – Corporate Research Laboratory formed | |
| | | **1964** – Ties up with GE to enter computer market. However, GE exits from mainframes in 1970 **1964** – Completes the remote control system for bullet trains |

| Critical events | Organizational features | Role of other firms |
| --- | --- | --- |
| | | **late 1960s** – Joint R&D with Japan National Railway to develop magnetically levitated trains |
| **1967** – Develops world's first automatic mail sorting equipment | | |
| | **1968** – Heavy Industrial Electrical Research Laboratory set up | **1968** – Completes world's largest electric power generation plant for Kansai Electric Power |
| **1970s** – Gains momentum in semiconductor chips; Medical Electronics also grows | **1970** – Production Engineering Laboratory set up | **1970s** – Develops CCDs (Charge Coupled Devices) for Canon, which put a stringent requirement for use in their autofocus cameras. CCDs are used by Toshiba's videocameras, photocopiers and facsimiles |
| **1970s** – Pioneers in the development of red, green, yellow gallium phosphide based LEDs for display indicators | **1970s** – Keen promoters of optoelectronics such as Shimora of the Corporate Research laboratory and Nagai, the director of Corporate Research laboratory | |
| **until 1970s** – White goods, TVs are main cash cow supporting the electronics and Office Automation businesses[3] | | **1970s** – Develops optical comm. system in the form of LAN, for Tokyo Electric Power and Chubu Electric Power[4] |
| **1972** – Launches the most powerful CMOS LSIs | | |
| **1974** – Completes ACOS mainframe computer | **1974** – Spins off Toshiba Chemical | |
| **1975–76** Begins research on semiconductor lasers for communications using GaAlAs lasers in the 0.8 μm region | | **1975** – Ties up with GE to develop BWR, next generation nuclear power system |
| | **1976** – After the oil shock, adopts GE's portfolio analysis to | |

| *Critical events* | *Organizational features* | *Role of other firms* |
| --- | --- | --- |

determine which
businesses they should
focus on. 300 business
units identified and
regrouped into 50;
Decides to focus on IT
and electronics which
are the major growth
areas

**1977** – Develops a
powerful CT scanner
**1978** – Plant in Nasu
for medical electronics
completed, costing 10
billion yen
**1978** – Decides to exit
from large mainframe
computer market and
to stay in the small
and medium sized
computer market for
distributed processing
systems[5]
**1978** – Develops
Japan's first Japanese
word processor

**1978** – 'Works Labs'
(laboratories within
the business units)
established, moving
about one third of the
workers from the
corporate research
laboratory to the
divisions, in the belief
that it is better to
separate long term
research from product
development
**1978** – Decides to
form a three-tier R&D
structure with
corporate level R&D
laboratories, business
sectoral level R&D
laboratories and the
Works laboratories
**1978** – Business units
were grouped into
sectors
**1978** – 'EPOC,
Engineering efficiency,
Productivity of
Coming age'
programme carried
out
**1978–9** Consumer
electronics research
laboratory, Nuclear
energy research
laboratory,

| *Critical events* | *Organizational features* | *Role of other firms* |
| --- | --- | --- |
| | Semiconductor laboratory set up | |
| **1979** – Merges Toshiba Trading which span off in 1953 | **1979** – Office Automation planning department set up, to promote sales of facsimiles, photocopiers, word processors etc. | **1979–1985** Takes part in the National Optoelectronics project. Works on semiconductor lasers which emit light of multiple wavelengths since it is possible to increase the amount of information transmitted |
| **1979** – World's first optical disk image data filing system, Tosfile developed (example of a Tokken type project), launched in 1981 and becomes the leader in the field | | |
| **1980** – Switches to Indium based semiconductor lasers in the 1.3 μm region; In parallel another group works on semicon. laser of short wavelength suitable for use in optical disk applications | **1980** – Establishes Materials business unit, merging the activities which had taken place in the separate business units. Medical Electronics Laboratory set up | **1980s** – Sets up joint venture with Motorola to develop microprocessors |
| **1980** – Research on LEDs completed | **1980** – 'POWER' programme carried out, to improve the efficiency of office workers | |
| **1980s** – President Saba is a keen promoter of globalization. Announces 'E & E' strategy, bringing together energy related fields and electronics such as transportation systems, OA, electric power generation, home electronics, through its competences in microelectronics, IT and materials[6] | | |

| Critical events | Organizational features | Role of other firms |
|---|---|---|
| **1980s** – Adopts a strategy of integrating MDS (Materials, Devices, Systems)<br>**1981** – Develops CMOS 64 K bit static RAM<br>**1982** – World's first gallium arsenide LSI developed | | **1982** – Two way technology licensing with Zilog for LSI and Microprocessor technology |
| | **1983** – New Materials promotion division set up | |
| **1984** – Announces 'I' Strategy of Information, Integration and Intelligence; Enters PBX (Private Branch Exchange) market | **1984** – VLSI research laboratory established<br>**1984** – Existing sectors are:-<br>Industrial electronics, Electronic components, Light electronics, Heavy industrial electronics, Materials. Each sector consists of several business units and a R&D laboratory | |
| **1985** – Begins mass production of 1 M DRAM chips for which Toshiba becomes the leading maker | | **1985** – Privatization of NTT. Until then, Toshiba was not a member of the NTT family. Opportunities to become a supplier for NTT arises from this point. Participates in NTT's plan to set up INS (Information, Network, System)<br>**1985** – Licenses 1 M DRAM chip technology to Siemens<br>**1985** – Joint development with SGS Thomson |

| Critical events | Organizational features | Role of other firms |
|---|---|---|
| | | Microelectronics to develop ICs for comm. **1985** – Joint development with LSI Logic to develop gate arrays |
| | **1986** – Begins a Toshiba group VAN service **1987** – Establishes System Software laboratory and Applied New Materials laboratory; Completion of the three tier R&D system which began in 1978 | |
| | **1988** – 4 Corporate level R&D labs, Corporate R&D lab, VLSI Research Centre, Manufacturing Eng. Centre, Systems Software lab exist; Employs 1750 people in the R&D labs. 8 Labs at the Sectoral level exist. 98% of R&D expenditure funded internally. R&D expenditure at the corporate laboratories funded by the business units accounts for 50% | **1988** – Sets up joint venture DTI (Display Technology Inc) with IBM to develop large screen colour TFT Liquid Crystal Displays |
| **1989** – IT and comm. account for over 50% of sales; The ratio of sales of semicon. laser for comm. use and IT use 50% each | | |
| | **1990** – Establishes LCD business unit | |

| Critical events | Organizational features | Role of other firms |
|---|---|---|
| | **1991** – Display Research lab set up (previously part of electronics component research lab) | |

[1]Toshiba Brochure, *Toshiba's History.*
[2]Mizobuchi (1990).
[3]*Toyo Keizai* (*Asian Economic Weekly*) *'Toshiba's Strategies'* (25 March 1989) (in Japanese).
[4]*Toyo Keizai* (*Asian Economic Weekly*), 'Toshiba's Strategies' (25 March 1989).
[5]Iwai (1984).
[6]Toshiba *Annual Report* (1989).

# Appendix E.10 Summary of Hitachi

| Critical events | Organizational features | Role of other agents |
|---|---|---|
| **1910** – Founded as an electrical repair shop for a copper mining company | **1910s** – Plants made profit centres; Product planning and marketing undertaken in the plants. Decentralized and divisionalized | **1910–40** Main customers are in the heavy industrial sector such as power, steel, and transportation. Licenses technology from GE on electric and nuclear power plants |
| **1910–40s** Company's main activities centred on ELM (Electro Locomotive Machinery).[1] Technology oriented company from its early days | | |
| **1924** – Completed the first large scale electric locomotive made in Japan | | |
| **1932** – Begins manufacturing elevators | | |
| | **1934** – Hitachi Research Laboratory founded. Founder president's motto "Though we cannot live 100 years, we should be concerned about 1000 years hence" | |
| | | **1940** – Completes an automatic telephone exchange for NTT, becoming part of the NTT family known as Denden family. Subsequently, Hitachi's relationship |

281

| Critical events | Organizational features | Role of other agents |
|---|---|---|
| | | with NTT centres on supplying exchanges, and not comm. systems |
| | **1942** – Central Research Laboratory | |
| **1950s** – Begins to diversify into consumer electrical market. Begins research on computers for internal use to control power and steel plants | founded | **1950s** – Licenses computer technology from RCA; Joins National project to develop high performance computers; Develops DIPS computer for data communication for NTT; Joint Development with Japan National Railway for railway reservation system |
| **1956** – Hitachi Cable and Hitachi Metal established | | |
| **1958** – Begins production of semiconductor chips | **1957** – Design Centre established | |
| **1959** – Completes electronic computers based on transistors | | |
| **1960s** – Attempts to balance imported technologies with in-house development | **1960** – Return on assets, Asset turnover, Profit Margins used as indices to assess the profitability of plants[2] | **1960s** – RCA working on LCDs |
| **1963** – Hitachi Chemical established | | |
| | | **1964** – Completes the first cars for bullet train (shinkansen) |
| **1965** – Begins mass production of colour TV tubes; TVs | | |

| *Critical events* | *Organizational features* | *Role of other agents* |
|---|---|---|

become a cash cow
feeding the
development of
computer and
semiconductor
business[3]

**1967** – Begins research
on III–V
semiconductors and
LEDs for numeral
indicators
**1969** – Begins research
on semiconductor
lasers for comm. in
the 0.83 μm region
(AlGaAs)
**1970s** – Strategy to
transform itself into
an electronics maker
**1970s** – Witnesses
computer sales
growth. Electronics
becomes the core
business domain
**1970** – Visible LED
transferred to
production but
business fails to grow
and withdraws from
production in 1973
(LEDs were taken
over by the arrival of
LCDs). However,
research continues on
infrared LED.
Begins research on
LCDs

**1966** – Mechanical
Engineering
Research Laboratory
founded[4]

**1970s** – Pioneers in
starting Tokken
(Special High Priority)
projects; 20% of R&D
expenditure spent on
Tokken (about 2000
people working on 20
projects)

**1971** – Energy
Research Laboratory,
Production Research
Laboratory founded[5]

| *Critical events* | *Organizational features* | *Role of other agents* |
|---|---|---|
| | **1973** – Systems Development Laboratory founded | |
| **1974** – Begins research on optical disks for computer storage | | **1974** – Commercial operation at Japan's first nuclear power plant starts |
| | **1975** – Device Devlopment Centre established within the computer business unit, to develop key components (LSIs) | **1975** – Collaborates with NTT to make telephone exchanges |
| **1976** – Begins research on DFB lasers | **1976** – Marketing managers begin to participate in the product development process | |
| **1977–8** Shifts to work on another material InP for semiconductor lasers; Work on AlGaAs semicon. lasers continued for other applications such as high power lasers for laser beam printers and optical storage | | |
| **1979** – Infrared LED commercialized | | |
| | | **1979–85** Participates in the National Optoelectronics Project; Works on OEICs |
| **1980** – Begins mass production of 64 K DRAM chips | **1980s** – R&D labs and plants get closer. Transfer of several hundred people a year from R&D to business units takes place. Optoelectronics Development promotion division set up | **1980s** – Benefits from strong demand for digital exchanges |
| **1980** – GaAlAs semicon. lasers commercialized for optical disks. Carried out as a Tokken project. Optoelectronics realized as a core technology | | |

| *Critical events* | *Organizational features* | *Role of other agents* |
|---|---|---|
| | **1981** – Mita takes over as president (has background in computers). Stresses the importance of business units and in some cases abolishes having plants as profit centres as in semiconductors | |
| **1982** – Hitachi's supercomputers launched<br>**1982** – InGaAsP lasers developed for communications in the 1.3 μm region | **1982** – OA (Office Automation) business unit set up | |
| **1984** – SAS structure laser developed<br>**1985** – Semiconductor lasers for laser beam printers, bar code readers launched | **1983** – Microelectronic Product Development lab established<br><br>**1985** – Advanced Research Lab set up<br>**1985–8** The number of people working on optoelectronics have grown 1.5 times | **1985** – Privatization of NTT. Since then NTT accepts bids from firms other than the NTT family, including ATT, Siemens and Toshiba |
| **1986** – Semiconductor lasers for CD pickup launched; Exits in 1988 due to severe competition<br>**1987** – DFB lasers commercialized | **1986** – Establishes marketing division within business units | **1986** – Joint research agreement with Bellcore of the US in the field of optical communication |
| | **1988** – Optoelectronics Research Laboratory set up in the Central Research Laboratory, bringing together the activities which were carried out in the separate units allocating 1/10 th of | |

| Critical events | Organizational features | Role of other agents |
|---|---|---|
| | resources at the CRL. At this stage, nine research labs exist including the Central Research Laboratory (ULSI, New Materials, Optoelectronics, Broadbank Network) and Hitachi Research Laboratory (Displays, Laser printers, Image Processing) | |
| **1989** – According to Fortune, Hitachi ranks third among the world's largest electronics firms rising from fifth in 1986 | **1989** – Semiconductor Design Development Centre established by separating the design unit within plants | **1989** – An agreement concluded with Texas Instrument to develop 16 M DRAM chips |
| **1989** – Semiconductor lasers for 16 G bit transmission developed. Suitable for HDTV transmission | **1989**–10 groups reporting to the president. Some large groups have several divisions (such as consumer products). Each group has several Works Labs | |
| **1990** – Sales of semiconductor laser for comm. equipment, optical disks amount to 50% each in value | **1990s** – R&D management bcomes an integral part of corporate planning | **1990** – Participates in the second national project on optoelectronics at Tsukuba |
| | **1990s** – Revision of the plant based profit centre approach | |
| | **1992** – A major reorganization to abolish plant based profit centred approach in most divisions[6] | |

[1]Shimizu (1989).
[2]Hitachi and its R&D', brochure (1991).
[3]Shimizu (1989).
[4]'Hitachi and its R&D', brochure (1989).
[5]EGIS report (1989).
[6]*Nikkei Business* (17 August, 1992).

# Appendix F   An Example of an INSPEC Record

# STN INTERNATIONAL®

INSPEC FILE SEARCH RESULTS – P207333J 27 JUL 90 01:23:13 PAGE 423

L29   ANSWER 446 OF 648
AN    84:2190372 INSPEC DN B84010086
TI    Mechanism of catastrophic degradation in InGaAsP/InP double-heterostructure light emitting diodes with Ti/Pt/Au electrodes.

AU    Ueda, O.; Yamakoshi, S.; Umebu, I.; Sanada, T.; Kotani, T. (Fujitsu Labs. Ltd., Atsugi, Japan)

SO    J. Appl. Phys. (USA) (Nov. 1983) vol.54, no.11; p. 6732–9; 7 refs.
      CODEN: JAPIAU ISSN: 0021-8979
      Price: CCC 0021-8979/83/116732-08$02.40
DT    Journal
TC    Experimental
LA    English

AB    The mechanism of catastrophic degradation of InGaAsP/InP double-heterostructure light emitting diodes with Ti/Pt/Au electrodes through the application of pulsed current, is investigated by electroluminescence topography, scanning electron microscopy, energy dispersive X-ray spectroscopy, and transmission electron microscopy. As the degradation is promoted, a dark region is generated in the center of the light emitting area. This then develops through the whole light emitting area. During the generation and the development process of the dark region, no dislocations are generated. The dark region, which develops through the whole light emitting area, consists of amorphous area and nonstructural small grains. There is clear compositional inhomogeneity in the region corresponding to the dark region. However, the electrode metals are not detected. The mechanism of the catastrophic degradation can be explained by local heating to the contact region by large current pulses, melting of the matrix crystal at the current above the degradation current level, and development of the molten region through the whole light emitting area.

CC    3B4260D
CT    ELECTROLUMINESCENCE; GALLIUM ARSENIDE; III–V SEMICONDUCTORS; INDIUM COMPOUNDS; LIGHT EMITTING DIODES; SCANNING ELECTRON MICROSCOPE EXAMINATION OF MATERIALS; TRANSMISSION ELECTRON MICROSCOPE EXAMINATION OF MATERIALS
ST    catastrophic degradation; InGaAsP/InP double-heterostructure light emitting diodes; Ti/Pt/Au electrodes; electroluminescence topography; scanning electron microscopy; energy dispersive X-ray spectroscopy; transmission electron microscopy; amorphous area; small grains; compositional inhomogeneity; dark region; local heating; large current pulses; melting
ET    As*Ga*In*P; As sy 4; sy 4; Ga sy 4; In sy 4; P sy 4; InGaAsP; In cp; cp; Ga cp; As cp; P cp; In*P; InP; Ti; V

\*\*\*\*\*\*\*\*\*\*\*\*\*\*\*\*\*\*\*\*\*\*\*\*\*\*\*\*\*\*\*\*\*\*\*\*\*\*\*\*\*\*\*\*\*\*\*\*\*\*\*\*\*\*\*\*\*\*\*\*\*\*\*\*

# Appendix G.1
# Questionnaire for the Firms Interviewed

The questions were divided into three parts as follows:

I.   General questions on the firm's optoelectronics-related R&D
IIa.  Questions related to the development of semiconductor lasers
IIb.  Questions related to the development of LCDs
III.  General questions on core technology management

Before the interviews took place, the questionnaire, my research outline and the analysis of the development of semiconductor lasers (and LCDs) derived from INSPEC were sent to the firms. At the interview, the results of the bibliometric analysis and technological linkage maps were shown.

## I. GENERAL QUESTIONS ON THE FIRM'S OPTOELECTRONICS-RELATED R&D

1. How and why did your firm enter the field of optoelectronics? Was it due to a top-down or bottom-up decision ?

2. How was the technology acquired or developed? (e.g. in-house effort, collaboration with other firms, national projects). What was the role of the government or the national PTT in helping the competence building ?
   What was the role of Universities?

3. On entering this new field, which existing techniques/technologies in your firm were particularly useful? What was the background of researchers?

4. How is optoelectronics perceived in your company? When and how did it come to be recognized as a core technology? What happened then?

5. Organizational chart. In which labs is optoelectronics-related R&D performed? How did the organization of optoelectronics-related R&D change over time?

6. To what extent does my analysis of the bibliometric INSPEC data and patent data (in 1976–89) reflect the state of your firm's optoelectronics-related R&D activities?. If it is not an accurate measure, where does the discrepancy lie, and how could it have arisen? To what extent can the concentration into sub-areas of optoelectronics as shown by such indicators be explained by the firm's accumulated technological and marketing competences? Are these targeted areas of concentration, stable

over time, or as the data shows are you likely to continue focusing on group X and Y? How does this relate to your firm's long-term strategy?

7. Self-assessment of the sub-areas of optoelectronics. (e.g. Optical communications, semiconductor lasers (for communications, for CD players, high power), photodiodes, optical disks, III–V semiconductors, OEIC, LCD, Optical fibres)

8. Please comment on my model of systems, key components and component generic technologies. What are your firm's strategies concerning each level? How can the integration between the three levels be made more efficiently?

9. Technological linkage maps have been produced by analyzing INSPEC data. Please confirm when research on the generic component technologies and the main key components listed started (and ended). Please state how important was the technology for developing the key component in the TYPE I map (Not Important, Important, Very Important) (the same for the TYPE II map).

10. Optoelectronics-related R&D expenditure over the last ten years.

11. Optoelectronics-related sales over the last 10 years, if possible, divided into the sub-areas of optoelectronics.

12. Variation of price of optoelectronic key components over time.

13. The variation of external sales (or internal consumption) ratio of optoelectronics key components over time.

14. Market share of optoelectronics-related key components.

15. The variation over time of the number of researchers working on the sub-areas of optoelectronics.

16. The technological life-cycle of the key components that are developed.

17. How is the integration between marketing, production and R&D achieved?

## IIA. QUESTIONS ON THE DEVELOPMENT OF SEMICONDUCTOR LASERS

1. My summary of the development of semiconductor lasers in your firm obtained from INSPEC data lists the important progress achieved. Have I missed out any key developments?

2. How did several trajectories emerge? How was the decision made to develop semiconductor lasers for different applications?

3. What was the main technological agenda of development of the different types of semiconductor lasers (for example, increasing life, increasing output power, decreasing the threshold current etc.)? How was the agenda set? Was it top-down, or due to trial and error of the researchers?

4. What are the appropriate parameters for measuring technological progress of semiconductor lasers?

5. Of the main developments summarised in my list, if there were some which did not lead to commercialization, what were the reasons?

6. Of the main developments which led to commercialization, please state which year it was commercialized.

## IIB. QUESTIONS ON THE DEVELOPMENT OF LCDS

1. My summary of the development of LCDs in your firm obtained from INSPEC data lists the important developments which took place. Have I missed out something?
2. The development of LCDs moved gradually over time from low end applications such as watches, to more sophisticated applications such as TVs and PCs. How did the trajectories evolve? LCD is an example of technological fusion. At each stage, what range of new technologies were developed? Which existing technologies were useful (for example IC technology)?
3. What was the main technological agenda of development of the different types of LCDs over time (Improving visibility, Faster speed of display)? How was the agenda set?
4. What are the appropriate parameters for measuring technological progress of LCDs?
5. Of the main developments summarized in my list, if there were some which did not lead to commercialization, what were the reasons?

## III. GENERAL QUESTIONS ON CORE TECHNOLOGY MANAGEMENT

1. Please comment on your firm's US patenting strategy, and the procedure for publishing papers.
2. What are the missions of the corporate R&D laboratories, application laboratories and the development units ?
3. The R&D project selection mechanism.
4. R&D funding mechanism; the percentage derived from the headquarter, business units and external sources.
5. What are your firm's features which could be considered as core competences (organizational, marketing, technological, R&D)? How are they different from the competitors'? How did your competences change over time? How have you been able to sustain competences?
6. Does your firm use the concept of core technologies?
   Which core technologies are you investing in?
7. Are technological investments made with a long-term view to develop technological capabilities and financial returns are of secondary importance, or are they subject to short-and medium-term financial constraints? Does it depend on the technology? If so, is there a mechanism to classify technologies? Do you use techniques such as TPM to assess technologies? How do you decide whether to develop a new technology in-house or to acquire it from external sources?
8. Are technological strategies formed independently of corporate strategies? Is it difficult to combine the two processes? What is the relative balance of power between accounting/finance people and technological people within your firm?

9. Is there a unit which is in charge of R&D (or technological) planning of the whole firm? What do they do exactly?
10. Is there a mechanism to integrate the know-how held by people in different divisions, units, functions (marketing, R&D etc.)? Do you have a system similar to Hitachi's Tokken? What are the merits of having such a system? How do you try to improve links between R&D and the business units?
11. When a new technology is developed, or acquired, what mechanism exists to diffuse that knowledge within the firm?
12. Do you have core technology champions? What do they do exactly?
13. Roughly, what percentage of top management have technical background?

# Appendix G.2 Individuals Interviewed

## CHAPTERS 1, 2 AND 3

Professor T. Ikoma, Institute of Industrial Science, Tokyo University

Professor Yamamoto, Electrical Engineering, Stanford University (jointly employed by NTT)

Dr. T. Sato, Executive Director, OITDA (Optoelectronics Industry and Technology Development Association)

Dr. W. Truscott, UMIST (University of Manchester, Institute for Science and Technology), Solid State Electronics Group

Professor T. Peaker, UMIST, Physics Department

Dr. B. Luff, University of Sussex, Physics Department

Mr. I. Molyneux, Chief Scientist, Pilkington Micronics

Dr. D. Jenkin, Product Development Manager, BT&D (British Telecom and Du Pont)

Mr. M. McLean, Scientific Generics, Cambridge

Dr. T. Raven, Scientific Generics, Cambridge

Mr. G. Mizoguchi, Scientific Generics (former staff of Canon)

Dr. U. Ito, Manager, Materials and Devices, R&D Centre, Eastman Kodak Japan (former staff of Eletrotechnical Laboratory, MITI)

Dr. K. Imanaka, Central R&D Laboratory, Omron, Kyoto

Mr. S. Saito, Recording Media Research Division, TDK

Mr. Y. Abe, Corporate R&D Dept, TDK

Mr. R. Moon, Project Manager, Optical Communication, Hewlett Packard, San Jose, USA

Ms. E. Wilson, European Manager, Corporate Manufacturing, Hewlett Packard, UK

Mr. T. Faulkner, Technical Assistant to the Director, Research, Eastman Kodak, Rochester, USA

Mr. H. Pollicove, Director, Center for Optics Manufacturing, University of Rochester, USA

Professor T. Yamanouchi, Faculty of Business Administration, Yokohama National University (former staff of Canon)

Dr. S. Tsukahara, Ministry of Education and Culture, Tokyo

Professor F. Kodama, Saitama University

Professor R. Nelson, Columbia University

Professor D. Okimoto, Asia Pacific Research Centre, Stanford University

Professor Y. Teramoto, Graduate School of Systems Management, Tsukuba University

Dr. T. Ray, PREST, University of Manchester

Dr. M. Boden, PREST, University of Manchester
Professor M. Teubal, Hebrew University of Jerusalem
Professor R. Hirasawa, Tokyo University
Professor M. Hirooka, Kobe University
Professor I. Dierickx, INSEAD
Professor Y. Doz, INSEAD
Professor P. Haspeslagh, INSEAD
Professor G. Dosi, University of Rome

## CHAPTER 4

Mr. N. Toyokura, General Manager of R&D Administration Division, Fujitsu Ltd.
Mr. S. Sakai, General Manager of R&D Administration Division, Fujitsu Ltd.
Mr. K. Asama, General Manager, R&D Planning and Coordination Office, Fujitsu Ltd.
Dr. H. Ishikawa, Semiconductor Laser Section, Optical Semiconductor Devices Lab, Fujitsu Laboratories
Dr. M. Sakaguchi, Vice President, Central Research Laboratory, NEC
Dr. K. Kobayashi, General Manager, Optoelectronics Research Laboratories, NEC
Mr. Y. Sakamura, Optical Cable Communication, Development Group, NEC
Dr. N. Shimasaki, Corporate Chief Engineer, NEC
Mr. M. Ugaji, Manager of R&D Coordination Department, Sony
Mr. K. Sakai, General Manager, R&D Coordination Department, Sony
Miss K. Kijima, R&D Corporate Planning Group, Sony
Dr. Y. Mori, General Manager, Optical and Functional Devices Research Department, Research Centre, Sony
Mr. Y. Ooki, Optoelectronics Research, Corporate Research Centre, Sony
Mr. T. Yoshida, Manager, Engineering Development, Audio Group, Sony
Mr. C. Kojima, General Manager, Compound Semiconductor Division, Sony
Mr. R. Yamayoshi, General Manager, Semiconductor Patent Division, Sony
Mr. K. Shintani, Deputy Manager, Components Development Group, Sony
Dr. Y. Kuwahara, Cooperate Senior Staff and Leader, Corporate R&D Promotion Division, Hitachi Ltd.
Dr. N. Chinone, Department Manager, Optoelectronics Department, Central Research Laboratory, Hitachi
Mr. Y. Nagae, Senior Researcher, Liquid Crystal Display section, Hitachi Research Laboratory
Dr. A. Kameoka, Senior Manager, Technical Planning, Toshiba
Dr. Nakamura, Research Planning, Corporate Research Center, Toshiba
Mr. S. Mori, Electronic Component Laboratory, Corporate Research Center, Toshiba
Mr. H. Ohashi, Electronic Component Laboratory, Corporate Research Center, Toshiba

Dr. Y. Naruse, Manager, Research Planning, Toshiba Corporation

Dr. H. Hayashi, Optoelectronics Research Lab, Sumitomo Electric

Mr. M. Koyama, General Manager of Optoelectronics R&D Lab, Sumitomo Electric

Dr. S. Kataoka, Director of Tokyo Research Lab, Sharp

Mr. Y. Nagaya, Technology Planning, Sharp

Dr. C. Bradley, Director of Sharp Europe, Sharp

Mr. M. Coupland, Optoelectronics Manager at STC Technology Ltd.

Mr. P. Selway, Optoelectronics Research Director, STC Technology Ltd.

Mr. G. Henshall, Senior Principal Research Engineer, STC Technology Ltd.

Dr. C. Hilsum, Director of Research, GEC Hirst Research Centre

Dr. W. Thulke, Optoelectronics Division, Corporate R&D, Siemens

Mr. R. Tobiasch, Corporate Research and Development Executive Office, Siemens

Mrs. B. Reminger, Manager, Technology Mapping Department, Siemens

Dr. K. Mettler, Senior Director, Optoelectronics Department, Siemens

Dr. S. Spiedel, Manager, Siemens (Tokyo)

Dr. S. Sudo, NTT Optoelectronics Laboratories

Mr. S. Yoshitake, Senior Manager, Engineering Strategy Planning Headquarters, NTT

Dr. Y. Itaya, Research Planning, NTT Optoelectronics Laboratories, NTT

Dr. J. Shimada, General Manager, Optoelectronics, Electrotechnical Laboratory, MITI

Dr. K. Tanaka, Director, Materials Science Division, Electrotechnical Laboratory, MITI


**CHAPTER 5**

Mr. D. Rossall, Online Marketing Officer of INSPEC, IEE


*Note*: The positions listed were held by the people when the author conducted interviews during October 1990–March 1992.

# Notes and References

## 1  Introduction

1. Prahalad and Hamel (1990); Teece, Pisano and Schuen (1990); Pavitt, (1989).
2. Craig Fields who is the Director of MCC (Microelectronics Computer Consortium, a government funded research institute in Texas) stressed that one must shift from the old model of a good company competing through good products, to one which can compete through building competences, at the conference on 'International Competitiveness in Optics and Imaging' organized by SPIE in Rochester, NY, (8–9 Oct. 1991).
3. A talk given by the Director of Research at Eastman Kodak, at the same conference in Rochester, NY (Oct. 8–9, 1991).
4. Penrose (1959); Nelson and Winter (1982); Teece, Pisano and Schuen (1990); Pavitt (1989); Dosi (1988).
5. We have divided optoelectronics into eight sub-fields.
6. See e.g. Stalk, Evans and Shulman (1991); Prahalad and Hamel (1990).
7. Imagine posing the question ten years ago to ask where companies 'A' and 'B' were going. If we were able to see that company 'A' would become a leader in ten years time, and 'B' the loser, then the management of 'B' might have been able to take some action.
8. Of course, predicting the future involves uncertainties which we can not guess.
9. Narin and Olivastro (1988).
10. E.g. M. Imai (1986).
11. Technological paths that a firm proceeds. We will discuss them further in Chapter 2.
12. A term proposed by Nelson and Winter (1977), organizational routines are similar to procedures in firms. Routinization of activity forms a way of storing organizational knowledge. We will discuss this further in Chapter 2.
13. Obviously if it is patented, it is private.
14. These component technologies are generic in nature, since they can be used to develop a range of components.
15. We will discuss this further in Chapter 3.
16. Such as life-time and performance parameters.
17. The most expensive ones for communications range between 100,000 to 500,000 yen, while those for compact disks may cost 300 yen (in 1990).
18. Data source: K. Miyazaki (1991). We will discuss this further in Chapter 3.
19. With the exception of Siemens who were cooperative.
20. With the exception of Philips, all the 11 finally chosen firms agreed to cooperate. In the early phase of my study I considered including some

US firms, but there were some financial constraints on the possibility of conducting fieldwork in the US. Nor have I included Matsushita or Canon. Thomson declined to cooperate and was excluded from the study.
21. UK, Germany and Japan.
22. Johnson (1982).
23. Abegglen and Stalk (1985).

## 2 Neo-Schumpeterian Theory

1. Machlup (1962).
2. Nelson and Winter (1977).
3. A distinctive culture which pervades in a company. Examples include ways of behaviour, common beliefs, aspirations, language and sets of values. See also Cremer (1990).
4. Teece, Pisano and Schuen (1990).
5. Factors affecting the selection process.
6. Porter (1980, 1985).
7. Including Oskarsson (1991); Granstrand (1991); and Kodama (1986b).
8. Mitchell (1986), p.227.
9. Mitchell (1986), p.230.
10. Metcalfe and Boden (1990), p.3.
11. E.g. Pavitt (1989); Doz (1989); Prahalad and Hamel (1990); Teece, Pisano and Schuen (1990).
12. Dierickx and Cool (1987).
13. The notion of path dependence includes a firm's previous investments and repertoire of routines which constrain its future behaviour (Teece, Pisano and Schuen, 1990)
14. These are assets which play a supportive but equally important role in a firm, such as distribution, after-sales support and marketing.
15. Author's interview with GEC (1990).
16. Shimizu (1989).
17. Author's interview at the Central Research Laboratory, Hitachi, (December 1991).
18. In supplier dominated sectors such as pulp, construction and publishing, most innovations come from the suppliers of equipment. They are generally small and their R&D capabilities weak. In production (scale) intensive firms producing standard materials, scale economies are strong. Technological leads are centred on process innovations. Specialized equipment suppliers provide their large customers with specialized knowledge and experience for building equipment for a variety of users. The technological trajectories are geared towards performance increasing product innovation.
19. Tushman and Anderson (1988).
20. Penrose (1959).
21. Narin and Olivastro (1988).
22. Into operation/investment/innovation by Westphal, Kim and Dahlman (1985).

23. Itami and Roehl (1987); Dierickx and Cool (1987).
24. Penrose (1959).
25. E.g. Giget (1984).
26. GEST (1984), p.36.
27. Teece (1982).
28. Myers (1984), p.135.
29. Hamilton and Mitchell (1990).
30. Stalk, Evans and Shulman (1992).
31. K-Mart used to be 'the King of the discount retailing industry' in 1979.
32. Cross-docking enables Wal-Mart to achieve the economies that come with purchasing full truck-loads of goods while avoiding the usual inventory and handling costs. This reduces Wal-Mart's costs of sales by 2 to 3% compared to industry average (Stalk *et al.*, 1992).
33. Doz, Prahalad, and Angelmar (1989).
34. Mitchell (1986).
35. Some which like nuts and bolts are cheap and abundant in supply, require little technical sophistication to produce and are basically mature products, adding low value added to the final products and can be thought of as commodities.
36. Different types of competences including technological competence are discussed earlier in section 2.3.
37. Nelson and Winter (1982).
38. Teece (1986).
39. E.g. Nelson and Winter (1982); Winter (1984).
40. E.g. Kay (1979).
41. Coombs *et al.* (1987), p.44.
42. Coombs *et al.* (1987), p.44.
43. Doz (1989).
44. Tushman and Anderson (1988).
45. Teece (1986).
46. Hobday (1994).
47. A similar decision was taken at Ampex and RCA which were the early pioneers.
48. Rosenbloom and Cusumano (1987).
49. E.g. Burns and Stalker (1961); Rothwell (1977); Freeman (1982); Souder (1983); Burgelman (1985).
50. Souder (1991).
51. Steiner (1965).
52. Pavitt (1986); Marengo (1991).
53. E.g. Learning by searching; Learning by training and hiring; Learning by designing and adaptation of product designs.
54. Dodgson (1991).
55. Arrow (1962); Hayes, Wheelwright and Clark (1988).
56. Rosenberg (1982); Nelson (1982); March (1991).
57. Itami and Roehl (1987).
58. Kline and Rosenberg (1986).
59. Hitachi Research Institute developed a technological evaluation technique in 1983, by categorizing technologies into 4 levels of system,

equipment (product), devices, and materials. Each level has four attributes of (1) science underlying the element, (2) realization of the scientific procedure, (3) production method, (4) links to lower level element (see Appendix A).

60.  However, we are fully aware that this pattern does not work in all sectors. There are companies (especially in the US) which have adopted a specialization route focusing on key components (e.g. Intel) and have succeeded. At the other extreme are some PC makers such as Dell, which have succeeded by assembling components sourced from other makers. (In the case of Dell, they use LCDs produced by Sharp, keyboards by Alps Electric, batteries by Sanyo and monitors by Sony.)

61.  A firm which also sells its components to external users.

62.  Author's conversation with Professor C. Freeman at SPRU.

63.  Doz (1989).

64.  Dodgson (1990).

65.  Rothwell *et al.* (1974); Schon (1973); Allen (1977).

66.  Project SAPPHO consists of the comparative analysis of 'paired' successful and unsuccessful innovations in the chemicals and scientific instruments industries. It examined the success/failure criteria (Rothwell, *et al.*, 1974).

67.  E.g. Myers and Marquis (1969); Langrish, Gibbons *et al.* (1972); Rothwell *et al.* (1974).

68.  E.g. Johnson (1982); Fransman (1990).

## 3   Optoelectronics – A Leading Core Technology

1.  The data are presented in yen, and have not been converted to US $, since the exchange rate appreciated from 238 yen to 128 yen during the period, and would have distorted the trend; see Appendix B.1 for detailed breakdown of these three broad categories.

2.  Interview by the author with Professor Ikoma, at Tokyo University, Institute of Industrial Science, March 1992.

3.  ACOST (1988).

4.  ACOST (1988).

5.  A quantum of electromagnetic radiation energy, proportional to the frequency of radiation.

6.  Mechatronics is the combination of mechanics and electronics.

7.  Blair (1991).

8.  Interview by the author with Dr. Chinone, Central Research Laboratory of Hitachi (1990).

9.  The amplitude of light that travels down the fibre gradually diminishes due to transmission losses.

10.  Over a kilometer of fibre, will give rise to a loss of 1000 dB (decibel).

11.  Hayashi, a Japanese scientist, later returned to Japan to work for NEC.

12.  Lee (1986), p.275.

13.  Senior and Ray (1990).

14.  Minimum absorption losses occurred. $\mu m = 10^{-6} m$

15. Absorption losses lead to the amplitude of the optical signal to diminish.
16. Interview by the author with Professor Y. Yamamoto, Professor of Electrical Engineering at Stanford University, who also holds a position at the NTT Basic Research Laboratory.
17. Optoelectronics key components are formed by forming very thin layers of materials of atomic dimension using epitaxial techniques. This abbreviation means that Gallium Aluminium Arsenide layer is formed on Gallium Arsenide.
18. Indium Gallium Arsenide Phosphide on Indium Phosphide.
19. See n. 16.
20. As automobiles are increasingly becoming sophisticated, fibreoptic technology is introduced for transmitting control data (e.g. engine control data).
21. Light Emitting Diode.
22. Kodama (1991).
23. Hilsum (1984).
24. ACOST (1988).
25. Interview by the author at GEC in 1990.
26. Involving simultaneous transmission of several messages along a single channel of communication.
27. TN displays could be driven by CMOS LSIs.
28. In recent years, TN LCDs have been declining in importance in Japan due to competition from manufacturers in the newly industrialized countries (NICS).
29. Other technologies for developing LCDs include TAB (Tape Automated Bonding) for packaging LSI directly onto a tape, and COG (Chip on Glass) for mounting LSIs onto glass. Making a thin panel gap of uniform thickness is an important technology in itself.
30. Interview by the author at Hitachi Research Laboratory, (December 1990).
31. ACOST (1988).
32. For example a 12 cm optical disk can store the equivalent of 1,000 floppy disks or 200,000 pages of A4 text (ACOST, 1988).
33. The figures for CD-ROM, optical WORM, 1 MB floppy disk and 40 MB hard disk are 0.5, 0.15, 0.1 and 0.05 seconds respectively.
34. The growth of a thin layer on a single crystal substrate that determines the lattice structure of the layer.
35. Uede (1980).
36. Into two dimensional form, thereby forming an image.
37. Gallium Aluminium Arsenide.
38. Indium Gallium Arsenide Phosphide.
39. In single mode, only one mode of light is emitted, compared to multimode, where several modes of light are emitted. In optical transmission, single mode fibres offer the advantage of being able to carry very large bandwidth.
40. Japan's Machinery Imports and Market, Opto-Electronics, No.1, 1988.

41. Solid lasers and gas lasers find their biggest applications in materials-processing such as cutting and welding, and in military applications.
42. Layers of different material.
43. A compound composed of elements from groups III and V of the periodic table (which often exhibits semiconductor characteristics).
44. Japan Machinery Importer's Association : Japan's Machinery Imports and Market, No. 1 (1988).
45. Author's interview at Hitachi, (March 1992).
46. Author's interview at Hitachi, see n. 45.
47. P Intrinsic type.
48. In multi-mode, different modes travel down the fibre with differing velocities.
49. Author's correspondence with Mike Coupland, former manager of optoelectronics research at STC (1993).
50. The companies involved were Sumitomo Electric, Furukawa Electric and Fujikura Electric.
51. Sudo (1982).
52. Fine glass particles, are deposited on a rotating target rod, to build up a preform layer by layer, (Sudo, 1982). Optical fibre is produced by drawing from the heated tip of the preform as it is lowered into a furnace (Wilson Hawkes, 1983).
53. According to Professor Yamamoto at Stanford University (and NTT), such optical fibres could be classified as one of the most significant breakthroughs of this century (interview by the author at Stanford, November 1992).
54. Goodhew and Humphreys (1988).
55. Lee (1986).
56. ACOST (1988).

## 4   Building Optoelectronics Competence In Firms

1. For a more historical breakdown please see Appendix D.
2. Post, Telephone and Telegraph company.
3. An arrangement of cartelising supply of telecommunications equipment for British Post Office between GEC, STC and Plessey.
4. Interview by the author at Sumitomo Electric (March 1992).
5. Interview by the author at Fujitsu (October 1990).
6. Wavelengths for microwaves range from 1mm to 10 cm or in other words $10^{-3}$ to $10^{-1}$m. Light waves are of the order of $10^{-6}$m. Light can carry 1,000 to 100,000 times more information than microwaves.
7. The Japanese companies were not, however, linked to defence.
8. Interview by the author at GEC (July 1990).
9. Although it was not possible to access Philips, its main business activities as shown in Table 4.1 point to Philips belonging to this group.
10. In Chapter 5, using the three types of data on publications, US patents, and interviews we measure competences in the sub-fields for all firms (ranging from 0 the weakest to 100 the strongest).
11. They also built competence in local area networks.

12. Toshiba's Annual Report (1991).
13. When the author returned to Japan in March 1992 to discuss the results of her analysis, the optoelectronics R&D manager agreed with the position of Hitachi as being evenly balanced between the two sectors.
14. Kobayashi (1985).
15. Ohmae (1984).
16. Kobayashi (1985).
17. Data on sales growth are shown in Appendix D.
18. Fujitsu Laboratory Report (1990).
19. Annual Report of NEC in (1975, 1980).
20. Although optical fibres might be classified as cables and wires, key components and systems would not.
21. Siemens Report by Vicker da Costa Ltd (1986).
22. Interview by the author at Siemens, Tokyo branch (November 1990).
23. Business Week, "Siemens Speeds Up", (20 February 1989), pp.16–20.
24. At NEC, for example, research on communications was conducted separately from that on electrical machinery until 1965.
25. Others include the basic research labaratory, microelectronics, C&C systems, and software engineering.
26. Interview by the author at STC, Harlow, July 1990.
27. Weinstock has been at the helm of GEC since 1963 and created a budget-driven style of management. (*Management Today* (January, 1989), p.41).
28. Interview by the author at GEC, Hirst Research Laboratory (July 1990).
29. Memo of Mick Mclean's interview with D.Roberts (27 June 1985).
30. The criticisms focused on its reliance on government contracts, on shedding its workforce while increasing its cash mountain and the failure to invest in new technologies (28 April 1985, House of Commons).
31. Chief executive of STC.
32. Similar to organizational procedures. They were discussed in Chapter 2.
33. Interview by the author at STC, Harlow (August 1991).
34. A method employed at GE, to analyze its portfolio of businesses.
35. Until then, components were made for internal use. A merchant maker sells components on the market.
36. Sharp's other areas of competences include microelectronics, solar energy, multi-media, mechatronics, digital magnetic storage and precision engineering.
37. Interview by the author at Hitachi, head office, R&D planning division (March 1992).
38. Interview by the author at Hitachi, Central Research Lab (November 1990).
39. See n. 37.
40. Interview by the author at Fujitsu (October 1990).
41. Nikkei Business (1992).
42. Interviews by the author at NEC and Toshiba (1990).
43. Mizobuchi (1990).
44. In the case of Sony, organizational charts do not apparently exist, because the organization changes so frequently (interview by the author with Sony, October 1990).

45. M. Miyazaki (1985).
46. Plessey's Caswell R&D centre had been working on semiconductor lasers.
47. Royal Signals and Radar Establishment.
48. Teece (1986).
49. Since the lasers had to be cooled with liquid helium the researchers lost confidence (interview by the author at Siemens, July 1991).
50. Organizational procedures or systems.
51. The Tokken system was pioneered at Hitachi.
52. For example, in the case of Sharp, TFT-type LCD TV, colour LCD for PC displays, LCD projection displays, magneto-optical disks, CCD and 1M DRAM were all carried out as Tokken projects (Trigger, Vol. 9, No. 10, 1990).
53. Ministry of International Trade and Industry.
54. Interview by the author with Dr Shimada, ETL (MITI's Electro Technical Laboratory) (November 1990).
55. See n. 54.
56. Sigurdson (1986); Fransman 1990; Abegglen and Stalk (1985).
57. See n. 54.
58. There were a number of other firms which participated, such as Furukawa Electric, Sumitomo Electric, Matsushita, Oki Electric and Shimadzu.
59. Shimada (1986).
60. See n. 54.
61. See n. 54.
62. V Channeled Substrate Inner Stripe Laser.
63. Interview by the author with Mike Coupland of STC (1992).
64. Senior and Ray (1990).
65. ACOST (1988).
66. ACOST (1988).

## 5 Empirical Results – The Classification of Competences

1. Over the last 50–60 years, the measurement and evaluation of scientific output using the publication and citation counting techniques of bibliometrics has gained widespread acceptance (Stephan and Levin, 1988), p.36. The publication indicator serves as an approximation for scientific activity and productivity (Weingart, Sehringer and Winterhager, 1988), p.396. Bibliometric techniques were developed by university groups and consultancy firms, to measure 'output' of research, and to map the development of new, emerging fields of S&T. Simple bibliometric analysis centres on counting scientific articles classified by authors/institutions or scientific fields, etc.
2. Nelson and Winter (1982).
3. Technometrics is an indicator developed by the Fraunhofer Institute to measure the competitiveness of R&D intensive products. The metric system is based on the various technical characteristics of products.

4. See e.g. Soete and Wyatt (1983), Narin and Davidson (1988), Pavitt and Patel (1988), Tunzelmann (1988), Schmoch and Grupp (1989).
5. In one company, the research manager said that sometimes papers are written even when the research fails.
6. For example, if the costs of filing for a patent outweigh the commercial or strategic benefits gained by filing.
7. Pavitt (1988b).
8. Pavitt (1988b).
9. In all seven Japanese firms studied, it was found that the patenting process was very similar.
10. III-V semiconductors include gallium arsenide materials, which are used in optoelectronic key components.
11. The online search was carried out through the Information Service section of the University of Sussex library.
12. More will be discussed later when the results of the empirical analysis is presented in section 5.3.
13. For example, we cannot compare GEC's self-assessment of 5 in liquid crystal displays with the score of 5 for Sharp, and conclude that they are equally strong. Nor can we say that judging from GEC's self-assessment of 4 in OEIC, and Toshiba's score of 3 that GEC is more competent than Toshiba in OEIC. These scores are all relative indices which are internal to the firm. Thus a company's score of 5 and 3 for communications systems and optical disks respectively indicates that they consider their competence to be high in communications systems and only average in optical disk systems.
14. Except for Sony where nine or ten people were interviewed.
15. The logarithmic version of RTA in international trade was first published by Wolter (1977), p.250–67. The logarithmic version is not good for distance or least squares measurements because of the poles (Grupp, 1993).
16. All the original INSPEC bibliometric data, IFTI data and other compiled data are available from the author.
17. All the original patent data, IFTI$_{Pat}$ data and other compiled data are available from the author.
18. A value of zero indicates the lack of involvement in the area.
19. Author's interview at STC (1991).
20. However since the mid-1980s the situation may have changed to some extent. The number of optoelectronics related patents granted in 1986–9 was almost triple that of 1977–81.

## 6 Learning, Interlinkages and Shifting Trajectories

1. In Oskarsson's (1991) study, Herfindahl indices were used to measure the increased technological diversification in Swedish Industries.
2. When company scientists were interviewed, they were asked to assess component generic technologies and key components according to this criteria.

3. This author's interview at Fujitsu (March 1992).
4. This author's interview. at Fujitsu (March 1992).
5. One person interviewed gave an example of how it happened to that firm in the field of semiconductor manufacturing equipment.
6. In the article 'Siemens Stresses the Importance of Components and Cooperation' in *Physics* World (April 1993), pp.17–18, it is mentioned that losses in Siemens' semiconductor business are rumoured to be 1 mDM a day.
7. Nelson's (1993) case of 'complementarity lock in' holds in optoelectronics.
8. Bloom (1992).
9. Since the systems team can develop the system knowing all the performance parameters while the key component is being developed, when the key component is sold on the open market, the firm would have had a lead time advantage. The firm which purchases the key component would have to start developing the system after buying it.
10. At Siemens, three trajectories of AlGaAs lasers emerged, one for laser beam printers, one for optical communication and a third one for sensors.
11. Such as those used at Heathrow airport.

## 7. Conclusions

1. E.g. Nelson and Winter (1982); Pavitt (1989); Doz (1989); Teece *et al.* (1990); Dosi (1988).
2. A field where the three-layer model would apply, having systems, key components and component generic technologies.

# Bibliography

J. C. ABEGGLEN, and G. STALK Jr (1985) *Kaisha, The Japanese Corporation* (New York: Basic Books).

ACOST (1988) *Optoelectronics: Building on Our Investment* (London: HMSO).

T. J. ALLEN (1970) 'Communications Networks in R&D Laboratories', *R&D Management*, 1, 14.

T. J. ALLEN (1977) *Managing the Flow of Technology* (Cambridge, MA.: MIT Press).

Z. AOYANAGI (1987) *Hikarigijutsu no Subetega Wakaru Hon* (my translation) *A Comprehensive Guide book to Optoelectronics Technologies*, (Kyoto: PHP Business Library).

K. J. ARROW (1962) 'The Economic Implications of Learning by Doing', *Review of Economic Studies*, 29: 155–173.

Y. BABA (1988) *Systemic Innovation: Its Nature and How to Benefit from it*, University of Sussex, SPRU, mimeo.

J. S. BAIN (1959) *Industrial Organization* (New York: John Wiley).

J. A. BLAIR (1991) *Proprietary Technology and Industrial Structure: The Semiconductor, Computer, and Consumer Electronics Industries*, A Special Report of the Northeast Asia–US Forum on International Policy, Stanford University.

M. BLOOM (1992) *Technological Change in the Korean Electronics Industry* (Paris: Development Centre, OECD).

R. BURGELMAN (1985) 'Managing the New Venture Division: Research Findings and Implications for Strategic Management', *Strategic Management Journal*, 6: 39–54.

R. BURGELMAN and R. ROSENBLOOM (1988) 'Technology Strategy: An Evolutionary Process Perspective', Chapter 1 in *Research on Technological Innovation, Management and Policy*, Vol. 4 (Greenwich, CN: JAI Press).

T. BURNS and G. STALKER (1961) *The Management of Innovation*, (London: Tavistock).

*Business Week* (1989) 'Siemens Speeds Up' (20 February): 16–20.

A. CAWSON, K. MORGAN, P. HOLMES, D. WEBBER and A. STEVEN (1987) *Hostile Brothers: Competition and Closure in the European Electronics Industry* (Oxford: Clarendon Press).

A. D. Chandler Jr (1962) *Strategy and Structure: Chapters in the History of Industrial Enterprise* (Cambridge MA.: MIT Press).

CICT (1989) K. Morgan, B. Harbor, M. Hobday, N. von Tunzelmann, W. Walker, The GEC–Siemens Bid for Plessey, SPRU *Working Paper* Series, No.2 (January).

W. M. COHEN and D. LEVINTHAL (1989) 'Innovation and Learning: The Two Faces of R&D', *The Economic Journal*, 99: 569–96.

K. COOL and D. SCHENDEL (1988) 'Performance Differences among Strategic Group Members', *Strategic Management Journal*, 9: 207–24.

R. COOMBS (1988) 'Technological Opportunities and Industrial Organisation' in Dosi *et al.* (eds) *Technical Change and Economic Theory*: 295–308.

305

R. COOMBS, P. SAVIOTTI and V. WALSH (1987) *Economics and Technological Change*, (London: Macmillan).

J. CREMER (1990) 'Common Knowledge and the Coordination of Economic Activities', in Aoki, *et al.* (eds) *The Firm as a Nexus of Treaties* (London: Sage): 53–75.

R. CYERT and J. MARCH (1963) *A Behavioral Theory of the Firm*, (Englewood Cliffs, NJ: Prentice-Hall).

I. DIERICKX and K. COOL (1987) *Competitive Advantage: A Resource based Perspective*, INSEAD paper (Fontainableau, France: INSEAD).

M. DODGSON (1990) *Strategy and Technological Learning: An Interdisciplinary Microstudy*, (paper prepared for the international conference Firm Strategy and Technological Change: Micro Economics or Micro Sociology organised by The Manchester School of Management, UMIST and Department of Economics, Manchester University, Manchester 27–28 September).

M. DODGSON (1991) 'Technology Learning, Technology Strategy and Competitive Pressures', *British Journal of Management*, 2: 133–49.

G. DOSI (1982) 'Technological Paradigms and Technological Trajectories', *Research Policy*, 11: 147–62.

G. DOSI (1984) *Technical Change and Industrial Transformation* (London: Macmillan).

G. DOSI (1988) 'Sources, Procedures and Microeconomic Effects of Innovation', *Journal of Economic Literature*, 26 (September): 1120–71.

G. DOSI and L. MARENGO (1992) *Some Elements of an Evolutionary Theory of Organizational Competences*, paper presented at the Conference of International Economic Association, Moscow (August).

Y. DOZ (1989) *Innovation, technologies and competencies:Mobilizing capabilities in companies*, (paper presented at OECD).

Y. DOZ, C. K. PRAHALAD and R. ANGELMAR (1989) 'Managing the Scope of Innovation: A Dilemma for Top Management', in R. Rosenbloom and R. Burgelman (eds) *Research on Technological Innovation Management and Policy*, 4, (Greenwich, CN: JAI Press).

G. DUNTEMAN (1989) *Principal Component Analysis, in Quantitative Applications in the Social Sciences* (London: SAGE University Papers, SAGE Publications).

EGIS (1989) *Japanese R&D Centres in Electronics 1988–89*, 22-1, Ichibancho, Chiyoda-ku,Tokyo.

*Electronic Business* (1989) 'Siemens Restructures R&D to be Closer to End Markets', (20 March).

M. FRANSMAN (1990) *The Market and Beyond, Cooperation and Competition in Information Technology in the Japanese System* (Cambridge: Cambridge University Press).

C. FREEMAN (1982) *The Economics of Industrial Innovation* (London: Pinter Publishers).

C. FREEMAN and C. PEREZ (1988) 'Structural Crises of Adjustment: Business Cycles and Investment Behaviour' in G. Dosi, C. Freeman, R. Nelson *et al.* (eds), *Technical Change and Economic Theory* (London: Pinter Publishers).

GEST (1984) *Grappes Technologiques: Les Nouvelles Strategies d'Entreprise*, (London: McGraw-Hill).

GIGET, M. (1984) Euroconsult, 'Les Bonzais de l'Industrie Japonaise', *CPE Etude* No. 40, (Paris).

P. J. GOODHEW and F.J. HUMPHREYS (1988) *Electron Microscopy and Analysis* (London: Taylor & Francis).

O. GRANSTRAND (1991) *The Economics of Multi-Tech: A Study of Multi-Technology Corporations in Japan, Sweden and the US*, paper presented for the international conference organized by the Science and Public Institute at Kaist, Seoul 30–31 October 1991.

O. GRANSTRAND and C. OSKARSSON (1991) *Technology Management in Multi-tech Corporations*, paper presented at the Portland International Conference on Management of Engineering and Technology – PICMET (27–31 October).

O. GRANSTRAND and S. SJOLANDER (1990) 'Managing Innovation in Multi-Technology Corporations', *Research Policy*, 19: 35–60.

O. GRANSTRAND, C. OSKARSSON, N. SJOBERG and S. SJOLANDER (1990) *Business Strategies for Development/Acquisition of New Technologies, A Comparison of Japan, Sweden and the US*, paper presented for the conference on Technology and Investment, arranged by the Royal Swedish Academy of Engineering Sciences (IVA) in cooperation with OECD and The Swedish Academy of Engineering Science, Stolkholm (21–24 January); later published as 'Business Strategies for New Technologies', in E. Deiaco, E. Hornell and G. Vrencery (eds), *Technology and Investment: Crucial issues for the 1990s* (London: Pinter) : 64–92.

Z. GRILICHES (1990) 'Patent Statistics as Economic Indicators: A Survey', *Journal of Economic Literature*, 28: 1661–1707.

H. GRUPP (1989) *Technology Indicators in Corporate Forecasting*, paper presented at the Colloquium on the Use of Technology Indicators in Strategic Planning, President Hotel, Brussels (23–24 January).

H. GRUPP (1994) 'The Measurement of Technical Performance of Innovations by Technometrics and Its Impact on Established Technology Indicators' *Research Policy*, 23, 2: 175–93.

H. GRUPP, U. SCHMOCH, B. SCHWITALLA and A. GRANBERG (1990) 'Developing Industrial Robot Technology in Sweden, West Germany, Japan and the USA' in J. Sigurdson (ed.) *Measuring the Dynamics of Technological Change* (London: Pinter.Publishers).

W. HAMILTON and G. MITCHELL (1990) 'What is Your R&D Worth?' *The McKinsey Quarterly*, No.3: 150–60.

R. H. HAYES, S. C. WHEELRIGHT, and K. CLARK (1988) *Dynamic Manufacturing Creating the Learning Organization*, (New York: Free Press).

B. HEDBERG (1981) 'How Organizations Learn and Unlearn' in P. Nystom and W. Starbuck (eds) *Handbook of Organizational Design: Volume 1* (Oxford: Oxford University Press).

D. HICKS, B. MARTIN and J. IRVINE (1986) 'Bibliometric Techniques for Monitoring Performance in Technologically Oriented Research: the Case of Integrated Optics' *R&D Management*, 16, 3: 211–23.

HIKARI SANGYO MAKER SOURAN (1987) (my translation) *Optical Industries' Makers Directory*, Sangyo Times, Tokyo.

C. HILSUM (1984) 'The Anatomy of a Discovery – biphenyl liquid crystals', in E. R. Howells (ed.), *Technology of Chemicals and Materials for Electronics* (Chichester: Horwood).

308                              *Bibliography*

HITACHI RESEARCH INSTITUTE (1983) *Electronics no Chouki Gijutsu Kaihatsu Senryaku,* (my translation) *Strategies for Technological Development in Electronics over the Long Term,* April, NIRA Report (NRC-81-3).

M. HOBDAY (1990) *Telecommunications in Developing Countries: the Challenge from Brazil* (London: Routledge).

M. HOBDAY (1994) 'Export-led Technological Development in the Four Dragons: the Case of Electronics', *Development and Change,* 25, 2 (April) 9: 333–61.

IEEE, INSPEC Marketing (1989) *Search Guide to Optics and Optoelectronics in the INSPEC Database.*

M. IMAI (1986) *Kaizen: The Key to Japan's Competitive Success* (New York: MacGraw-Hill).

K. IMAI and Y. BABA (1989) *Systemic Innovation and Cross Border Networks,* paper presented at the International Seminar on the Contributions of Science and Technology to Economic Growth at the OECD, Paris, (June).

J. IRVINE and B. Martin (1989) *Research Foresight* (London: Pinter Publishers).

ITAMI, H. and T. ROEHL (1987) *Mobilizing Invisible Assets* (Boston: Harvard University Press).

M. IWAI (1984) *Toshiba no Sentaku, Hachi Dai Project o Ugokasu E&E Senryaku* (my translation) *Toshiba's Choice: E&E Strategy for Managing 8 Large Scale Projects,* (Tokyo: Diamond Publishers,).

JAPAN MANAGEMENT ASSOCIATION (Nippon Nouritsu Kyokai) (1987) *Senshin Kigyou 30 Sha ni Miru Gijutsusha Kyouiku JitsuRei Shu* (my translation) *The Case Studies on Training of Scientists and Technologists in 30 Major Firms,* (Tokyo).

C. JOHNSON (1982) *MITI and the Japanese Miracle: the Growth of Industrial Policy in 1925–75* (Stanford, CA: Stanford University Press).

A. JOJIMA (1990) *Sharp Kaihatsu Saizensen* (my translation) *Sharp's R&D at the Frontier* (Tokyo: Sekai Bunkasha).

N. KANDEL, T. DURAND, J. REMY and C. STEIN (1991) 'Who's who in Technology: Identifying Technological Competence within the Firm', *R&D Management,* 21,3.

K. C. KAO and G. A. HOCKHAM (1966) 'Dielectric-fiber Surface Waveguides for Optical Frequencies', *Proc IEE,* 133, 7: 1151–58.

J. KATZ (1985) 'Domestic Technological Innovations and Dynamic Comparative Advantages:Further Reflections on a Comparative Case-Study Program', in N. Rosenberg and C. Frischtak (eds), *International Technology Transfer* (New York: Praeger).

N. KAY (1979) *The Innovating Firm* (London: Macmillan).

S. J. KLINE and N. ROSENBERG (1986) 'An Overview of Innovation', in R. Landau and N. Rosenberg (eds), *The Positive Sum Strategy: Harnessing Technology for Economic Growth* (Washington, DC: National Academy of Sciences Press).

K. KOBAYASHI (1985) *Computers and Communications* (Boston: MIT Press).

F. KODAMA (1986a) 'Japanese Innovation in Mechatronics Technology', *Science and Public Policy,* (February): 44–51.

F. KODAMA (1986b) 'Technological Diversification of Japanese Industry', *Science,* 233, (18 July).

F. KODAMA (1991) *Analyzing Japanese High Technologies* (London: Pinter Publishers).

F. KODAMA (1992) 'Technology Fusion and the New R&D', *Harvard Business Review*, (July–August): 70–8.

T. KOJIMA and K. IKEHATA (1984) *Nippon Denki to Fujitsu* (my translation) *NEC and Fujitsu* (Tokyo: Nippon Jitsugyou Publishers).

J. LANGRISH, M. GIBBONS, W. EVANS and F. JEVONS (1972) *Wealth from Knowledge* (London: Macmillan).

D. L. LEE (1986) *Electromagnetic Principles of Integrated Optics* (London: John Wiley).

F. MACHLUP (1962) *The Production and Distribution of Knowledge in the United States*.

J. G. MARCH (1991) 'Exploration and Exploitation in Organizational Learning', *Organization Science*, 2, 1: 71–87.

L. MARENGO (1991) *Knowledge Coordination and Learning in an Adaptive Model of the Firm*, DPhil Thesis, Science Policy Research Unit, University of Sussex, (September).

E. MASON (1949) 'The Current State of the Monopoly Problem in the US', *Harvard Law Review*, (June).

S. J. METCALFE and M. BODEN (1990) *Strategy, Paradigm and Evolutionary Change*, paper for the workshop 'Processes of Knowledge Accumulation and the Formulation of Technology Strategy', Rosnaes, Zealand, Denmark, (20–23 May).

G. MITCHELL (1986) 'New Approaches for the Strategic Management of Technology', in M. Horwitch (ed) *Technology in the Modern Corporation, a strategic perspective* (Oxford: Pergamon Press): 132–44.

K. MIYAZAKI (1993) *The Dynamics of Competence Building in European and Japanese Firms: the Case of Optoelectronics*, DPhil Thesis, University of Sussex, (June).

M. MIYAZAKI (1985) *Sumitomo Denkou: Kigyou nai Venture*, (my translation) *Sumitomo Electric: Intra-firm Ventures* (Tokyo: Kanki Publishers).

Y. MIZOBUCHI (1990) *Hikari Disk Konshinkai no Ayumi to Shourai* (my translation) *The Progress and Future of the Optical Disk Discussion Group*, OITDA, Optoelectronic Industry and Technology Development Association of Japan, (May).

A. MORITA (1987) *Made in Japan* (London: Fontana/Collins).

D. MOWERY (1983) 'Industrial Research and Firm Size. Survival and Growth in American Manufacturing, 1921–1946: An Assessment', *Journal of Economic History*, 43: 953–80.

D. MOWERY and N. ROSENBERG (1979) *The Influence of Market Demand upon Innovation: A Critical Review of Some Recent Empirical Studies* (Stanford, CA: Stanford University).

S. MYERS and D. G. MARQUIS (1969) *Successful Industrial Innovation: a Study of Factors Underlying Innovation in Selected Firms*, National Science Foundation, Washington, DC, NSF 69–17.

MYERS, S. (1984) 'Finance Theory and Finance Strategy', *Interfaces*, 14, 1: 126–37.

F. NARIN, E. NOMA and R. PERRY (1985) *Patent Based Indicators of Corporate Technological Strength*, Washington, DC: CHI Research, George Washington University.

F. NARIN and J. DAVIDSON (1988) *The Growth of Japanese Science and Technology: Science and Technology Indicators Show a Dramatic Strengthening*, Washington, DC: CHI Research, George Washington University.

F. NARIN and D. OLIVASTRO (1988) 'Technology Indicators based on Patents and Patent Citations', in A. F. J. van Raan (ed) *Handbook of Quantitative Studies of Science and Technology* (Elsevier Science Publishers B. V., North-Holland) ch. 15.

R. NELSON (1982) 'The Role of Knowledge in R&D Efficiency' *The Quarterly Journal of Economics*, (August).

R. NELSON (1991) 'The Role of Firm Differences in an Evolutionary Theory of Technical Advance', SPRU 25th Anniversary Papers, *Science and Public Policy*, 18, 6, (December: 347–52).

R. NELSON (1993) 'Co-Evolution of Technologies and Institutions' (seminar, SPRU, University of Sussex).

R. NELSON and R. LEVIN (1986) 'The Influence of Science, University Research and Technical Societies on Industrial R&D and Technical Advance', Policy Discussion Paper Series 3. Yale University, Research Program on Technology Change, (New Haven, CN).

R. NELSON and S. WINTER (1973) *'Neoclassical vs. Evolutionary Theories of Economic Growth: Critique and Prospectus'*, Mimeo.

R. NELSON and S. WINTER (1977) 'In Search of a Useful Theory of Innovation', *Research Policy*, 6: 36–76.

R. NELSON and S. WINTER (1982) *An Evolutionary Theory of Economic Change* (Cambridge, MA: Belknap Press).

*Nikkei Business* (1992) 'Hitachi Seisakusho Fukkatsu no Jouken' (my translation) 'The Conditions of Hitachi's Recovery', (17 August) 10–25.

OECD (1989) *Science and Technology Policy Outlook 1988*, (Paris: OECD).

K. OHMAE (1984) *Excellent Company – Nippon Denki no Sougou Kenkyu (my translation) Excellent Company – A Comprehensive Analysis of NEC* (Tokyo: President-sha).

OITDA (1990) (Optoelectronic Industry and Technology Development Association of Japan) *OPTONEWS*, 56.

C. OSKARSSON (1991) *Technology Diversification – The Phenomenon, Its Causes and Effects,* (Goteborg, Sweden: Chalmers University of Technology).

K. PAVITT (1984) 'Sectoral Patterns of Technical Change: Towards a Taxonomy and a Theory' *Research Policy*, 13, 9: 343–73.

K. PAVITT (1986) 'Foreword', in G. Bertin and S. Wyatt, (eds) *Multinationals and Industrial Property: The Control of the World Technology*, (Harvester, Wheatsheaf): 11–20.

K. PAVITT (1988a) A Critique of Tushman and Anderson's chapter in A. Pettigrew (ed.), *The Management of Strategic Change* (Oxford: Blackwell): 123–7.

K. PAVITT (1988b) 'Uses and Abuses of Patent Statistics' in A. F. J. van Raan (ed.) *Handbook of Quantititive Studies of Science and Technology* (North-Holland: Elsevier Science Publishers): 509–36.

K. PAVITT (1989) 'Strategic Management of the Innovating Firm', in R. Mansfield (ed.) *Frontiers of Management* (London: Routledge).

K. PAVITT (1991) 'Key characteristics of the Large Innovating Firm', *The British Journal of Management* London: 2: 41–50.

K. PAVITT and P. PATEL (1988) 'The International Distribution and Determinants of Technological Activities', *Oxford Review of Economic Policy*, 4: 35–55.

K. PAVITT and P. PATEL (1991) 'Technological Strategies of the World's Largest Companies', *Science and Public Policy*, (December) 363–68.

K. PAVITT, M. ROBSON, and J. TOWNSEND (1989) 'Technological accumulation, diversification and organization in UK companies 1945–83', *Management Science*, 35, 1 (January): 81–99.

E. PENROSE (1959) *The Theory of the Growth of the Firm* (Oxford: Blackwell).

K. POLANYI, (1962) Personal Knowledge, (New York: Harper).

M. E. PORTER (1980) *Competitive Strategy: Techniques for Analyzing Industries and Competitors* (New York: Free Press).

M. E. PORTER (1985) *Competitive Advantage: Creating and Sustaining Superior Performance* (New York: Free Press).

C. K. PRAHALAD and G. HAMEL (1990) 'The Core Competence of the Corporation', *Harvard Business Review*, (May–June): 79–91.

N. ROSENBERG (1976) *Perspectives on Technology* (Cambridge: Cambridge University Press).

N. ROSENBERG (1982) *Inside the Black Box: Technology and Economics* (Cambridge: Cambridge University Press)

R. ROSENBLOOM and M. CUSUMANO (1987) 'Technological Pioneering and Competitive Advantage: The Birth of the VCR Industry', *California Management Review*, 29, 4, (Summer): 51–76.

R. ROTHWELL (1975) 'Intracorporate Entrepreneurs', *Management Decision*, 13, 3: 142–54.

R. ROTHWELL (1977) 'The Characteristics of Successful Innovators and Technologically Progressive Firms', *R&D Management*, 3, 3: 191–206.

R. ROTHWELL (1991) *The Fifth Generation Innovation Process*, paper presented at the conference on Anregung und Erfolgscrientierte Durchsetzung von Innovation, Munich, (9–10 October).

R. ROTHWELL (1992) 'Successful Industrial Innovation: Critical Factors for the 1990s', *R&D Management*, 22, 3: 221–39.

R. ROTHWELL *et al.* (1974) 'Sappho Updated – Project Sappho Phase II', *Research Policy*, 3, 3: 258–91.

U. SCHMOCH and H. GRUPP (1989) 'Patents between Corporate Strategy and Technology Output: An Approach to the Synoptic Evaluation of US, European and German Patent Data', in A. F. J van Raan, A. J. Nederhof and H. F. Moed (eds), *Science Indicators: Their Use in Science Policy and their Role in Science Studies*, (Leiden: DWSO Press): 49–67.

D. A. SCHON (1973) 'Champions for Radical New Inventions', *Harvard Business Review*, (March–April).

J. M. SENIOR and T. RAY (1990) 'Optical-fibre Communications: the formation of Technological Strategies in the UK and USA', *International Journal of Technology Management*, 5, 1: 71–88.

C. SHAPIRO (1985) 'Patent Licensing and R&D Rivalry', *American Economic Review Papers and Proceedings*, (May).

C. SHAPIRO (1989) 'The Theory and Business Strategy', *Rand Journal of Economics*, (Spring).

J. SHIMADA (1986) *Memories on Optoelectronics Project with Love and Sorrow*, Electrotechnical Laboratory, MITI, mimeo.

K. SHIMIZU (1989) *Hitachi DaiHenbou* (my translation) *The Great Transformation of Hitachi* (Tokyo: TBS Britannica).

G. SIEMENS (1957) *History of the House of Siemens* (Munich: Karl Aber Freiburg).

J. SIGURDSON (1986) *Industry and State Partnership in Japan. The Very Large Scale Integrated (VLSI) Project*, Research Policy Institute, (Lund, Sweden).

L. SOETE and S. M. E. WYATT (1983) 'The Use of Foreign Patenting as an Internationally Comparable Science and Technology Output Indicator', *Scientometrics*, 5: 31–54.

W. E. SOUDER (1991) 'Organizing for Modern Technology and Innovation: A New Review and Synthesis' *Issues in the Management of Innovation*: 61–75.

W. E. SOUDER (1991) 'Organizing for Modern Technology and Innovation: A New Review and Synthesis' in Rosegger (ed,) *Management of Technological Change*, rev. edn (Oxford: Elsevier Advanced Tech.): 61–75.

G. STALK, P. EVANS and L. SHULMAN (1992) 'Competing on Capabilities: The New Rules of Corporate Strategy', *Harvard Business Review*, (March–April): 57–69.

G. STEINER (1965) *The Creative Organization*,: proceedings of a Seminar Sponsored by the Graduate School of Business in 1962, (Chicago University Press).

P. E. STEPHAN and S. G. LEVIN (1988), 'Measures of Scientific Output and the Age–Productivity Relationship', in A.F.J van Raan (ed.) *Handbook of Quantitative Studies of Science and Technology* (North-Holland: Elsevier).

E. STERNBERG (1992) *Photonic Technology and Industrial Policy* (New York: State University of New York Press).

S. SUDO (1982) *Studies on the Vapor-Phase Axial Deposition Method*, PhD Thesis, Tokyo University (July).

D. J. TEECE (1980) 'Economies of Scope and Scope of the Enterprise', *Journal of Economic Behaviour and Organization*, 1: 223–47.

D. J. TEECE (1982) 'Towards an Economic Theory of the Multiproduct Firm', *Journal of Economic Behaviour and Organization*, 3: 39–63.

D. J. TEECE (1986) 'Profiting from Technological Innovation: Implications for Integration, Collaboration, Licensing and Public Policy', *Research Policy*, 15: 285–305.

D. J. TEECE (1988), 'Technological Change and the Nature of the Firm' in G. Dosi *et al.* (eds), *Technical Change and Economic Theory* (London: Pinter Publishers).

D. TEECE, G. PISANO and A. SCHUEN (1990), *Firm Capabilities, Resources, and the Concept of Strategy*, CCC Working Paper, No. 90–8 (University of California, Berkeley).

N. VON TUNZELMANN (1988) *Convergence of Firms in Information and Communication: a test using Patent Data* (SPRU, University of Sussex), mimeo.

M. TUSHMAN and P. ANDERSON (1988), 'Technological Discontinuities and Organization Environments', in A. Pettigrew (ed.), *The Management of Strategic Change* (Oxford: Blackwell): 89–122.

*Trigger* (1990) *Sharp KisoKenkyu no Vision* (my translation) *Vision of Sharp's Basic Research*, 9, 10: 6–28.

B. TWISS (1974) *Managing Technological Innovation* (London: Longman).

M. UENOHARA (1987) 'Nippon Denki no Restructuring' (my translation) 'NEC's Restructuring', *The Journal of Science Policy and Research Management*, 2, 1: 34–9.

VICKERS DA COSTA (1986) *Review of Siemens Report*, (London) (January).

P. WEINGART, R. SEHRINGER and M. WINTERHAGER (1988), 'Bibliometric Indicators for Assessing Strengths and Weaknesses of West German Science', in A. F. J van Raan (ed.) *Handbook of Quantitative Studies of Science and Technology* (North-Holland: Elsevier).

L. WESTPHAL, L. KIM, and C. DAHLMAN (1985), 'Reflections on the Republic of Korea's Acquisition of Technological Capability', in N. Rosenberg and C. Frischtak (eds). *International Technology Transfer* (New York: Praeger): 167–221.

O. WILLIAMSON (1975) *Markets and Hierarchies: Analysis and Antitrust Implications* (New York: Free Press).

S. G. WINTER (1987) 'Knowledge and Competence as Strategic Assets', in D. J. Teece, *The Competitive Challenge: Strategies for Industrial Innovation and Renewal* (Center for Research in Management, Berkeley, CA).

WOLTER, B. (1977) 'Factor Proportions, Technology and West-German Industries International Trade Patents', *Weltwirtschaftliches Archiv*, Band 113: 250–67.

# Index

Note: 'n.' after a page reference indicates the number of a note on that page.